METALLIC GLASSES - PROPERTIES AND PROCESSING

Edited by **Hu Huang**

Metallic Glasses - Properties and Processing
http://dx.doi.org/10.5772/intechopen.72335
Edited by Hu Huang

Contributors

Wei Guo, Shusen Wu, Shulin Lü, Abdenacer Berradja, Hongwei Zhao, Dan Zhao, Bo Zhu, Shunbo Wang, Ning Li, Hu Huang

First published in London, United Kingdom, 2018 by IntechOpen
IntechOpen is the global imprint of INTECHOPEN LIMITED, registered in England and Wales, registration number: 11086078, The Shard, 25th floor, 32 London Bridge Street
London, SE19SG – United Kingdom
Printed in Croatia

British Library Cataloguing-in-Publication Data
A catalogue record for this book is available from the British Library

Additional hard copies can be obtained from orders@intechopen.com

Metallic Glasses - Properties and Processing, Edited by Hu Huang
p. cm.
Print ISBN 978-1-78923-720-7
Online ISBN 978-1-78923-721-4

For EU product safety concerns:
IN TECH d.o.o., Prolaz Marije Krucifikse Kozulić 3, 51000 Rijeka, Croatia,
info@intechopen.com or visit our website at intechopen.com.

Meet the editor

Dr. Hu Huang received his B.E. and Ph.D. degrees in Mechanical Engineering from Jilin University, China, in 2010 and 2014, respectively. Supported by the Japan Society for the Promotion of Science, he worked as a research fellow at Keio University from 2014 to 2017. Currently, he is a full professor at Jilin University. His research interests include precision instruments, mechanical properties of materials (metallic glass, hard-brittle materials, etc.), and surface modification of materials. He has published more than 50 peer-reviewed journal papers, which were indexed by the Science Citation Index Expanded (Web of Science).

Contents

Preface

Metallic glasses with long-range disordered atomic structure possess unique mechanical, chemical, and physical properties compared to their crystalline counterparts, and they are regarded as very promising engineering and functional materials, attracting increasing attention from both scientific and industrial communities. Although great progress has been achieved for metallic glasses in the last two decades, three main problems, including dimensional limit, poor tension plasticity, and difficulty in machining and shaping, still hugely hinder their wider applications as practical structural and functional materials. Thus, further investigation of the preparation, properties, and processing of metallic glasses is urgently required. This book provides its readers with recent achievements and developments of these aspects, and six chapters are included.

Chapter 1 is the introductory chapter, which explains the nature and purpose of the book, as well as its contents.

Chapter 2 reviews some pivotal aspects during the thermoplastic forming of metallic glasses, including mainly the evaluation of thermoplastic formability and its relationship with material flowing characteristic, thermoplastic forming techniques for processing metallic glass components with high quality, and the potential applications of these thermoplastic-formed textures.

Chapter 3 reviews recent developments in the structure of metallic glasses and their mechanical properties and deformation behaviors investigated by atomic-level simulations.

Chapter 4 focuses on the fabrication and mechanical properties of metallic glass matrix composites, which has been confirmed to be a significant method for improving the room-temperature plasticity of metallic glasses.

Chapters 5 and 6 focus on the tribo-electrochemical aspect of metallic glasses, including basic knowledge, tribocorrosion, passivity breakdown, and tribological and corrosion behaviors.

Several colleagues have contributed to this work, and I am very grateful for their help in its completion. This book is expected to assist students and researchers in their work, and contribute to new developments in the field of metallic glasses.

Hu Huang
Jilin University, China

Section 1

Introduction

Chapter 1

Introductory Chapter: Properties and Processing of Metallic Glasses

Hu Huang

Additional information is available at the end of the chapter

http://dx.doi.org/10.5772/intechopen.78665

1. Introduction

With an amorphous atomic structure, metallic glasses (MGs) (also called amorphous alloys) own some unique features compared to the conventional metal alloys, which make them versatile materials [1–7]. For example, MGs commonly show very high strength [8, 9], and thus they are very promising materials for fabrication of aircraft frames. High hardness and excellent resistance to wear make them potential candidates for contact applications such as phone's shell. The high elasticity makes them suitable for applications as golf clubs or spring [10]. High-strength, low elastic modulus, and good corrosion resistance make MGs promising applications as biomedical materials [2]. Although great progress has been achieved for MGs in the past decades, their practical applications as structural and functional materials are greatly impeded due to three main problems [1, 11], that is, dimensional limit, poor tension plasticity, and hard-to-machining and shaping.

As the formation of amorphous structure requires a high cooling rate, the size of MGs is greatly limited. Although some MGs with relatively good glass forming ability (GFA) are developed recently, the size of MG sample is quite small compared to that of conventional metal alloys. Currently, some new methods such as laser additive manufacturing are employed to produce MGs [12–15], which are promising to solve the problem of dimensional limit. However, additional issues such as crystallization and crack appear, which require further investigation.

For applications as structural materials, tensile plasticity is necessary. However, most MGs exhibit brittle fracture under tensile stress due to the fast propagation of single or very few shear bands [16, 17]. Therefore, improving the tensile plasticity of MGs is a very significant topic for their practical applications. Recently, some methods such as introduction of secondary phase into the glass matrix (i.e., fabrication of metallic glass matrix composites (MGMC)) [18–20], mechanical pre-deformation [21], surface modification [22–25] are proposed to tune the tensile plasticity of MGs, achieving some progress. However, the foundational mechanism and physical nature for plastic deformation and improvement of the plasticity of MGs are still not very clear. Further investigation by advanced experiments and simulations is required.

For practical applications of MGs, shaping them into parts is the first step. Due to the superplasticity within the supercooled liquid region (SCLR), thermoplastic forming provides a unique method for processing MGs, which has been widely employed and investigated [4, 26, 27]. However, the flexibility of this method is quite low, and this method also requires expensive molds. Furthermore, it may be only suitable for MGs with large SCLR; otherwise, crystallization may occur. Apart from thermoplastic forming, conventional machining such as diamond cutting [28–32], and non-conventional machinings such as micro-electrical discharge machining (micro-EDM) and laser irradiation [33–36] are also attempted for processing MGs. Some progress is achieved by these methods but also challenges remain [29, 36, 37]. Further investigations on these processing methods as well as developing new methods are required.

2. The structure of this book

According to the description in Section 1, to solve the problems that hinder the practical applications of MGs, further investigation on preparation, properties, and processing of MGs is urgently required. This book aims to present some recent achievements and developments on properties and processing of MGs, mainly including thermoplastic forming of metallic glasses, atomic-level simulation on mechanical deformation of metallic glasses, metallic glass matrix composites, as well as tribo-electrochemical applications of metallic glasses.

Author details

Hu Huang

Address all correspondence to: huanghu@jlu.edu.cn

School of Mechanical Science and Engineering, Jilin University, Changchun, Jilin, China

References

[1] Plummer J, Johnson WL. Is metallic glass poised to come of age? Nature Materials. 2015;**14**:553-555

[2] Li HF, Zheng YF. Recent advances in bulk metallic glasses for biomedical applications. Acta Biomaterialia. 2016;**36**:1-20

[3] Marti K. Sampling the sun. Science. 2007;**318**:401-402

[4] Kumar G, Tang HX, Schroers J. Nanomoulding with amorphous metals. Nature. 2009; **457**:868-872

[5] Schroers J. Bulk metallic glasses. Physics Today. 2013;**66**:32-37

[6] Inoue A, Nishiyama N. New bulk metallic glasses for applications as magnetic-sensing, chemical, and structural materials. MRS Bulletin. 2007;**32**:651-658

[7] Inoue A, Takeuchi A. Recent development and applications of bulk glassy alloys. Inter national Journal of Applied Glass Science. 2010;**1**:273-295

[8] Amiya K, Inoue A. Fe-(Cr,Mo)-(C,B)-Tm bulk metallic glasses with high strength and high glass-forming ability. Materials Transactions. 2006;**47**:1615-1618

[9] Amiya K, Urata A, Nishiyama N, Inoue A. Fe-B-Si-Nb bulk metallic glasses with high strength above 4000 MPa and distinct plastic elongation. Materials Transactions. 2004; **45**:1214-1218

[10] Telford M. The case for bulk metallic glass. Materials Today. 2004;**7**:36-43

[11] Hofmann DC, Suh JY, Wiest A, Duan G, Lind ML, Demetriou MD, Johnson WL. Designing metallic glass matrix composites with high toughness and tensile ductility. Nature. 2008;**451**:1085-10U3

[12] Shen YY, Li YQ, Chen C, Tsai HL. 3D printing of large, complex metallic glass structures. Materials and Design. 2017;**117**:213-222

[13] Yang C, Zhang C, Xing W, Liu L. 3D printing of Zr-based bulk metallic glasses with complex geometries and enhanced catalytic properties. Intermetallics. 2018;**94**:22-28

[14] Li N, Zhang JJ, Xing W, Ouyang D, Liu L. 3D printing of Fe-based bulk metallic glass composites with combined high strength and fracture toughness. Materials and Design. 2018;**143**:285-296

[15] Ouyang D, Li N, Xing W, Zhang JJ, Liu L. 3D printing of crack-free high strength Zr-based bulk metallic glass composite by selective laser melting. Intermetallics. 2017;**90**:128-134

[16] Sun BA, Wang WH. The fracture of bulk metallic glasses. Progress in Materials Science. 2015;**74**:211-307

[17] Xu J, Ramamurty U, Ma E. The fracture toughness of bulk metallic glasses. JOM. 2010; **62**:10-18

[18] Ning ZL, Liang WZ, Zhang MX, Li ZZ, Sun HC, Liu AL, Sun JF. High tensile plasticity and strength of a CuZr-based bulk metallic glass composite. Materials and Design. 2016;**90**:145-150

[19] Hays CC, Kim CP, Johnson WL. Microstructure controlled shear band pattern formation and enhanced plasticity of bulk metallic glasses containing in situ formed ductile phase dendrite dispersions. Physical Review Letters. 2000;**84**:2901-2904

[20] Qiao JW, Jia HL, Liaw PK. Metallic glass matrix composites. Materials Science and Engineering Reports. 2016;**100**:1-69

[21] Gu J, Song M, Ni S, Liao XZ, Guo SF. Improving the plasticity of bulk metallic glasses via pre-compression below the yield stress. Materials Science and Engineering A. 2014;**602**:68-76

[22] Fu J, Zhu YH, Zheng C, Liu R, Ji Z. Effect of laser shock peening on the compressive deformation and plastic behavior of Zr-based bulk metallic glass. Optics and Lasers in Engineering. 2016;**86**:53-61

[23] Qu RT, Zhang QS, Zhang ZF. Achieving macroscopic tensile plasticity of monolithic bulk metallic glass by surface treatment. Scripta Materialia. 2013;**68**:845-848

[24] Cao YF, Xie X, Antonaglia J, Winiarski B, Wang G, Shin YC, Withers PJ, Dahmen KA, Liaw PK. Laser shock peening on Zr-based bulk metallic glass and its effect on plasticity: Experiment and modeling. Scientific Reports. 2015;**5**:10789

[25] Zhang Y, Wang WH, Greer AL. Making metallic glasses plastic by control of residual stress. Nature Materials. 2006;**5**:857-860

[26] Li N, Chen W, Liu L. Thermoplastic micro-forming of bulk metallic glasses: A review. JOM. 2016;**68**:1246-1261

[27] Schroers J. Processing of bulk metallic glass. Advanced Materials. 2010;**22**:1566-1597

[28] Bakkal M, Shih AJ, Scattergood RO. Chip formation, cutting forces, and tool wear in turning of Zr-based bulk metallic glass. International Journal of Machine Tools & Manufacture. 2004;**44**:915-925

[29] Bakkal M, Liu CT, Watkins TR, Scattergood RO, Shih AJ. Oxidation and crystallization of Zr-based bulk metallic glass due to machining. Intermetallics. 2004;**12**:195-204

[30] Bakkal M, Shih AJ, Scattergood RO, Liu CT. Machining of a Zr-Ti-Al-Cu-Ni metallic glass. Scripta Materialia. 2004;**50**:583-588

[31] Bakkal M, Shih AJ, McSpadden SB, Liu CT, Scattergood RO. Light emission, chip morphology, and burr formation in drilling the bulk metallic glass. International Journal of Machine Tools & Manufacture. 2005;**45**:741-752

[32] Fujita K, Morishita Y, Nishiyama N, Kimura H, Inoue A. Cutting characteristics of bulk metallic glass. Materials Transactions. 2005;**46**:2856-2863

[33] Huang H, Yan JW. Surface patterning of Zr-based metallic glass by laser irradiation induced selective thermoplastic extrusion in nitrogen gas. Journal of Micromechanics and Microengineering. 2017;**27**:075007

[34] Huang H, Jun N, Jiang MQ, Ryoko M, Yan JW. Nanosecond pulsed laser irradiation induced hierarchical micro/nanostructures on Zr-based metallic glass substrate. Materials and Design. 2016;**109**:153-161

[35] Huang H, Yan JW. On the surface characteristics of a Zr-based bulk metallic glass processed by microelectrical discharge machining. Applied Surface Science. 2015;**355**:1306-1315

[36] Huang H, Yan JW. Microstructural changes of Zr-based metallic glass during microelectrical discharge machining and grinding by a sintered diamond tool. Journal of Alloys and Compounds. 2016;**688**:14-21

[37] Huang H, Noguchi J, Yan JW. Shield gas induced cracks during nanosecond-pulsed laser irradiation of Zr-based metallic glass. Applied Physics A: Materials Science and Processing. 2016;**122**:881

Section 2

Thermoplastic Forming

Chapter 2

Thermoplastic Forming of Metallic Glasses

Ning Li and Jiang Ma

Additional information is available at the end of the chapter

http://dx.doi.org/10.5772/intechopen.78016

Abstract

Metallic glasses (MGs) are an unusual class of materials that possess an amorphous atomic-level structure and display a plethora of desirable mechanical, chemical and physical properties, which makes them one of the most promising engineering materials. However, the poor processability of metallic glasses greatly hindered their engineering applications. Though some techniques have been developed to fabricate metallic glass components, the unique superplasticity of supercooled liquid metallic glasses attracts enduring attentions, which allows thermoplastic forming of metallic glasses on length scales ranging from atomic-size to centimeter and especially offers an alluring prospect in the field of microfabrication. While some pivotal aspects during thermoplastic forming of metallic glasses should be addressed, for example, the evaluation of thermoplastic formability and its relationship with material flowing characteristic, the required thermoplastic forming techniques for processing MG components with high quality and the potential applications of these thermoplastic formed textures are compressively reviewed in this chapter.

Keywords: metallic glasses, thermoplastic forming, formability, supercooled liquid region, vibration loading, superhydrophobic

1. Introduction

Unlike crystalline metals where dislocations or grain boundaries carry the plastic deformation, metallic glasses (MGs) usually deform inhomogeneous plastic deformation at ambient temperature caused by high localization of shear stress, resulting in fail catastrophe with zero tensile plasticity [1], which severely constraints their structural applications in macro-scale. This challenge tends to be mediated by reducing the sample size or feature below a critical length scale (<1 mm), wherein large tensile-plasticity and enhanced strength could be observed [2, 3], exhibiting size-dependent deformation behavior. Furthermore, MGs also illustrate size-dependent crystallization kinetics at nano-scale, such as the crystallization temperature rapidly increases with reduction in the diameter of nanorods, disclosing the enhanced thermal stability [4]. Consequently, the potential applications of MGs in micro- and nano-fields such as micro and nano-electro mechanical systems (MEMS/NEMS) have attracted enduring attentions [5]. However, the poor manufacturing ability origins from the high strength and ambient-temperature brittleness has been the Achilles' heel to structural applications of MGs [6, 7]. In the past decade, efforts have been devoted to fabricate MGs components with precise and versatile geometries, though the main techniques mainly focus on mold casting [8], thermoplastic forming [5, 8–26] and additive manufacturing [27–29]. By comparing with mold casting and additive manufacturing, the superiorities of thermoplastic forming is worth noting, for example, (1) the existence of supercooled liquid regime (SCLR) between the glass-transition temperature (Tg) and the crystallization temperature (Tx) allows thermoplastic forming (TPF) of MGs under low-forming strength [6], which breaks through the limitations of poor processability of MGs at ambient temperature; (2) net-shaping of precise and versatile geometries with minimum size of atom-scale could be realized, that were previously unachievable with any conventional crystalline metals; (3) the absence of phase transition of MGs during solidification endows them small solidification shrinkage (1/20 of typical casting alloys) [30], which is beneficial to the net-shaping with high precision and (4)as mentioned earlier, MGs maintain more excellent mechanical properties than crystalline metals.

In investigating the thermoplastic micro-forming of MGs, formability, namely the filling ability of supercooled liquid MGs in the mold, has been proposed to the MGs processability in the supercooled liquid region [31]. For MGs with various alloy compositions, previous literatures have reported that the thermoplastic formability was related to fragility of the supercooled liquid MGs and the width of supercooled liquid region. While for an MG with certain composition, the low viscosity and the long processing time are always appreciated [8, 32], in which the viscosity of supercooled liquid MGs is determined by processing parameters such as temperature, stress and strain rate [33]. The forming parameters actually affect the materials flow characteristics (i.e. Newtonian and non-Newtonian flow) [34]; therefore, the fundamental understanding the correlation between materials flow characteristics and thermoplastic formability is attractive with great significance. To improve the thermoplastic formability of supercooled liquid MGs, various forming techniques have been developed; these novel methods could also hot-process MGs components with macro size. It is worth noting that the potential applications of these thermoplastic formed parts especially the micro-components/patterns have been probed, which would broaden the real application of MGs. On the basis of the above descriptions, this chapter reviews the related aspects and provides in-depth understanding of the fundamental issues.

2. Thermoplastic formability

In order to evaluate the filling ability of supercooled liquid MG, micro-nano imprinting experiments on geometrical transferability of V-grooved die shapes to the material was first carried out by Saotome et al. [35, 36], who regarded that the micro-nano formability of supercooled liquid MGs could be quantified by the percentage of flowed area (R_f), expressed as,

$$R_f = A_f / A_g \quad (1)$$

where, A_f is the flowed area into the V-groove, A_g is the area of the V-groove. They found that the supercooled liquid MGs exhibit superior formability on micro-nanometer scales. It is easy to find that the alloys used in thermoplastic forming are generally with wide temperature ranges of supercooled liquid region, $\Delta T_x = T_x - T_g$, to reduce the risk of crystallization. Large ΔT_x indicates that the MGs have opportunities to obtain low viscosity, long forming time, and enhance the thermoplastic formability. Accordingly, ΔT_x is also regarded as one of the important indicators of the formability [31, 37], similar to the normalized parameter (S) [30] that should reflect better the formability of a MG, particularly when comparing different MG alloy families,

$$S = \Delta T_x / (T_l - T_g) \quad (2)$$

in which T_l is the melting temperature. While for MGs with different compositions, fragility parameter (m) [38],

$$m = \partial \log\eta / \partial (T_g / T) \big|_{T=T_g} \quad (3)$$

was proposed to measure the formability of supercooled liquid MG [37]. As the Angell plots of conventional MGs and high entropy MG as shown in **Figure 1**, wherein the temperature dependent viscosity among alloys exhibits various steepness index, that is, fragility parameter (m). A large steepness index corresponds to fragile liquid behavior, such as $Pt_{57.5}Cu_{14.7}Ni_{5.3}P_{22.5}$ MG shows the largest value of fragility, exhibits fragile liquids with the best micro-formability, is ideal candidate for near-net shape processing with fine printability. While a small index corresponds to strong liquid and exhibits poor formability, such as the thermoplastic forming of TiZrHfNiCuBe, high entropy MG becomes arduous with reducing mold size to tens micrometer, owing to the strong supercooled TiZrHfNiCuBe high entropy MG with small value of m [39]. Similar results are also observed by Schroers [31], who proposed a simple and precise standard to characterize the formability of BMGs, and the maximum diameter (d) of the hot-formed disc was taken as a measure of the MG's formability.

It is essential that all these parameters (such as R_f, ΔT_x, m and d) in evaluating the thermoplastic formability of MGs focus on amorphous alloys with various compositions. As for MG with certain composition, it is well understood that low viscosity is crucial to improve the thermoplastic formability of supercooled liquid MG. The viscosity of MGs not only depends on the temperature but is also sensitive to the strain rate. For example, with increasing strain rates under a certain temperature in the supercooled liquid region, there is a remarkable decrease in the viscosity, accompanied by the transitioning from Newtonian to non-Newtonian behavior [40]. In this case, thermoplastic forming becomes increasingly difficult, rather than enhancement [41]. In order to probe the physical origin of this phenomenon, Li et al. [34] established a

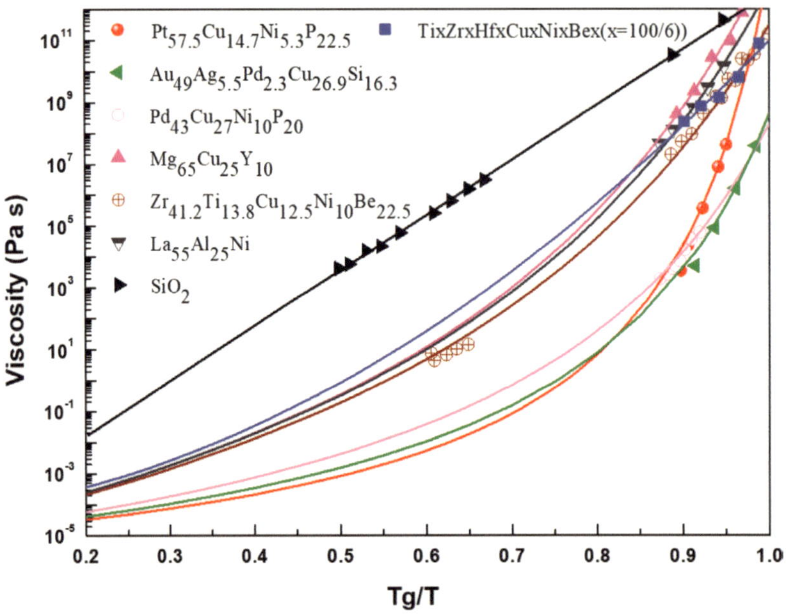

Figure 1. Angell plots of conventional MGs and high entropy MG [39].

thermoplastic forming map (see **Figure 2**), which reveals an inherent relationship between the thermoplastic formability and the flow characteristics, namely, Newtonian flow facilitates the forming capability, while the thermoplastic forming in a non-Newtonian flow regime tends to be difficult. Li et al. believe that this scenario is caused by the spatio-temporally homogeneous/inhomogeneous flow of MGs in Newtonian/non-Newtonian flow regime.

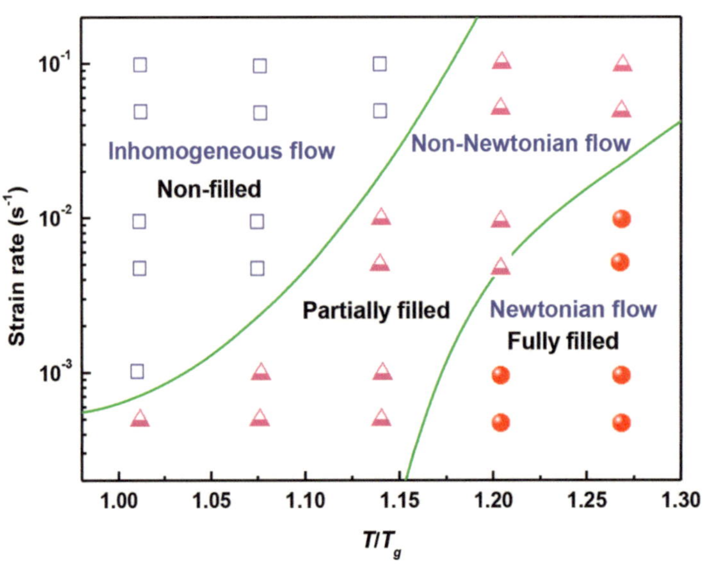

Figure 2. Thermoplastic forming map that reveals the relationship between the formability and the flow characteristics [34].

http://dx.doi.org/10.5772/intechopen.78016

3. Thermoplastic forming techniques

Thermoplastic forming map clarifies the relationship between flow features and formability and provides the selection of processing parameters. However, the Newtonian flow usually locates at regions with high processing temperature and low strain rate, which would induce the crystallization of amorphous alloys. In addition, the interfacial effect between amorphous alloys and mold materials becomes prominent during micro- and nano-scale forming, which seriously hinders the forming of metallic glasses [5, 9]. In order to improve the formability of supercooled MGs, various forming techniques have been developed.

By comparison with the hot-embossing technique as mentioned earlier, injection molding [42] as a net-shaping method for MGs exhibits superiorities in development of commercial manufacturing processes with minimized production cycle and high-volume production. Wherein the feedstock melt is gathered and forced into the part forming mold cavity at high pressure and velocity. As a potential forming process for MG parts, the injection molding is conducted at temperatures much lower than direct casting, which can improve

Figure 3. These shapes were previously unachievable with any other metal processing method that can be fabricated by blow molding [46].

the lifetime of the mold. Furthermore, the processing is accomplished in the laminar flow regime; therefore, higher quality and reliable parts could be obtained by comparison with the current mold-casting technique [8, 42]. However, the viscosity of the supercooled liquid MGs is much higher than that of the plastics melt, which poses a challenge for mass production.

In order to improve the thermoplastic formability of supercooled liquid MGs, micro-back-extrusion was proposed by Wu et al. [14], and a three-dimensional cup-shaped object with wall thickness of 0.05 mm was successfully fabricated. To reduce the contact area between MGs and mold materials, rolling was developed by Schroers et al. [43] who not only hot-rolled high-quality MG sheets but also replicated micro-patterns with featured size of 300 nm. The micro-replication of MGs through hot-rolling is actually similar to hot-embossing process, wherein the high viscosity and interfacial effect are main reasons limit the processability. Subsequently, Schroers et al. [44, 45] developed blow molding (see **Figure 3**), which allows blowing hollow products by using gas pressure to inflate the thermoplastic MGs enclosed in the mold. The low-forming pressure and high-dimensional accuracy indicates that this net-shaping technology could bring economic and environmental benefits.

Recently, an ultra-fast MGs' hot-processing technique was probed by Johnson et al. [47], as illustrated in **Figure 4**. When rapidly and uniformly heating a metallic glass at rates of 10^6 K/s to temperatures spanning the undercooled liquid region, rapid thermoplastic forming of the undercooled liquid into complex net shapes is implemented under rheological conditions typically used in molding of plastics. Owing to the millisecond time window, this method is able to "beat" the intervening crystallization and successfully process even marginal glass-forming alloys with very limited stability against crystallization that are not processable by

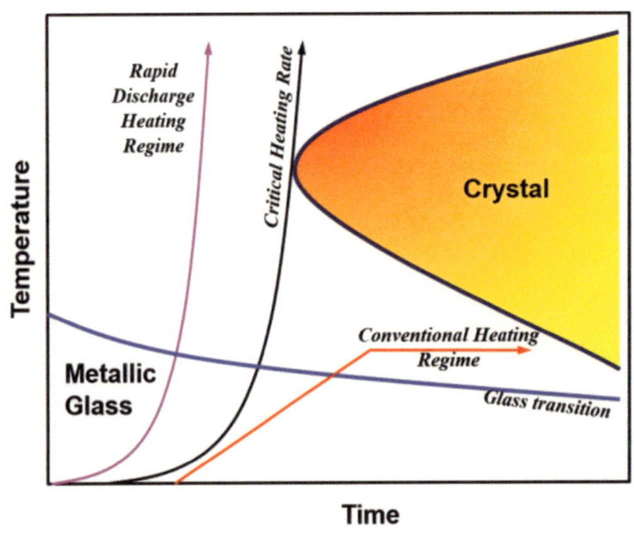

Figure 4. Using the rapid uniform heating approach with heating rates in the order of 10^6 K/s, the undercooled liquid is accessible at any temperature above the glass transition, through the melting point and beyond, where the liquid enters the equilibrium state.

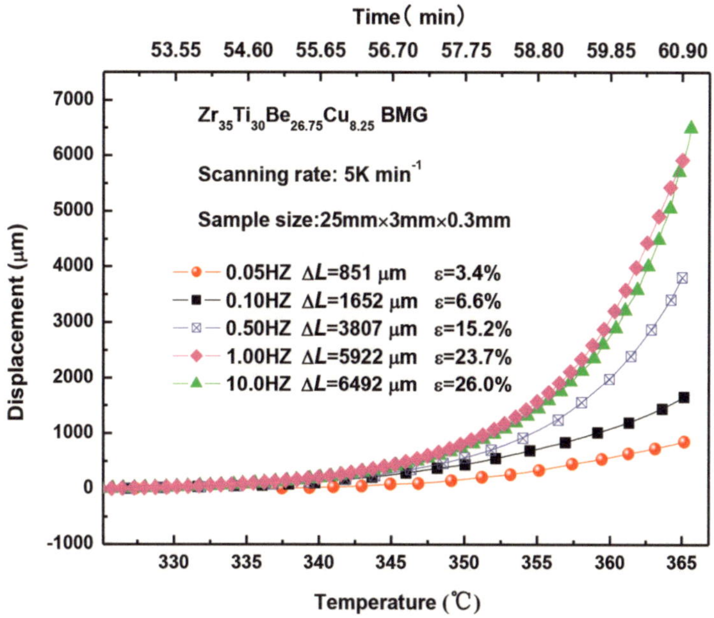

Figure 5. The displacement-temperature (time) curves of $Zr_{35}Ti_{30}Be_{26.75}Cu_{8.25}$ MG after vibrational tension under various loading frequencies ranging from 0.05 to 10 Hz (temperature rises from 23 to 365°C with scanning rate of 5°C/min^{-1}) [50].

conventional heating. Take advantage of unique rheological property along with the classic Lorentz force concept, electromagnetic coupling of electric current and a magnetic field was then thermoplastically shape a metallic glass without conventional heating sources or applied mechanical forces [48].

Based on improvements of formability made in the traditional metal formed by employing ultrasonic vibration [49], and considering that the viscosity is closely related to the dynamic relaxation of the alloy system, namely the shortening of the relaxation time, reduced viscosity is caused. Li et al. [50] introduced vibrational loading in thermoplastic forming of MGs; the intriguing finding was that the formability of supercooled liquid MGs is facilitated by vibrational loading (**Figure 5**). This technique exhibits potential applications in micro-/nano-scale forming of MGs. By increasing loading frequency to about 20 KHz, Ma et al. [51] used high frequency ultrasonic beating method to fabricate micro- to macro-scale structures, avoiding crystallization and oxidation of MGs.

4. Potential applications

The above thermoplastic forming techniques endow MGs with superiority in net-shaping precise and versatile structures comprising of macro-/micro-/nano-sized features. Through nanoimprinting, Schroers et al. [5, 8, 9] fabricated metallic glass nanowires with very high aspect ratios (>200); these nanorods not only exhibit enhanced thermal stability [4] but also display superb durability combined with high electrocatalytic activity toward methanol, ethanol oxidation and

CO, exhibiting great potential in energy conversion/storage and sensors fields [52]. The superb durability and high-surface area of these MG nano-structures motivate the generation of first functional proton exchange membrane micro fuel cells (MFCs). Such novel MFCs have been identified as a promising alternative power sources for portable electronics [53].

In addition to the potential applications in energy sector, the micro-/nano-gratings hot-imprinted on MGs surfaces also exhibit excellent spectroscopic performance [54, 55]. For example, Chu et al. [54] fabricated nano-scale gratings, and Ma et al. [55] hot-embossed micro-scale gratings with fine periodicity on Pd-based MGs surfaces, both surface exhibit beautiful optical properties such as rainbow-like spectrum when shone by fluorescent lamp light, as shown in **Figure 6**. Inoue et al. [56] pointed out that these nano-imprinted MG surfaces exhibit potential applications as anti-reflection materials, electrode materials, hologram technology, next generation ultra-high density of information data storage material and cell culture medium for bio-chips.

By integrating macro-, micro- and nano-scale features in a sequential order, Kumar et al. [13] hot-embossed hierarchical structures and displayed potential applications in optical devices, electrochemical activity and cellular response. Through micro-imprinting, some micro-lens arrays [57], micro-channel geometries [58] have been fabricated, showing potential applications in aspheric lens and fuel cell interconnect plates, respectively. Furthermore, the thermoplastic formed MG components have been used as a master mold (see **Figure 7**) to imprint polymers (such as PMMA) [10, 24, 59, 60], and an integrated PMMA micro-channel part was fabricated, implying that MG is a robust, attractive and viable mold material for thermoplastic imprinting of polymer devices [10]. Bardt et al. [23] thermoplastic formed some complex 3-D micro-topologies and envisaged potential application as high-Q micro-resonators, microwave waveguides, microsurgical tools and devices, connectors for higher frequency operations, micro-scale motors and transmission components, microfluidic arrays, and free-form reflective micro-optics.

The hot-embossed surface micro-components can be used in MEMS, biochips, such as micro-spring, micro-gear, micro-motor, micro-fan, micro-honeycomb structure, micro-gyroscope and micro-accelerometer structure and micro-turbines; some beautiful surface features such as micro-bats and micro-poetry of Tang Dynasty "Yellow Crane Tower" have also been fabricated

Figure 6. (a) Photographs of polished BMG plate (left) and BMG grating (right) when fluorescent lamp light shines upon them (b) photographs of Si die (left) and BMG grating (right) under the shine of fluorescent lamp light [55].

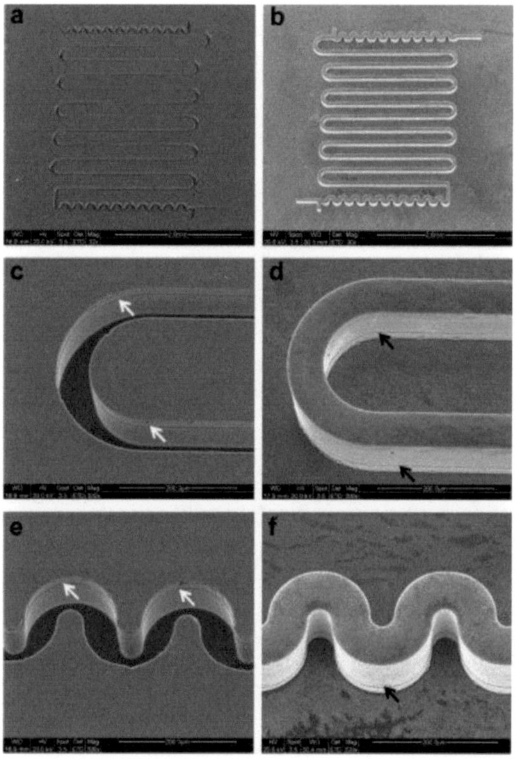

Figure 7. SEM micrographs of the silicon master (a) with single (c) and continuous bends structure (e); the corresponding hot-embossed metallic glass micro-channel structure (b, d, and f) [10].

by Li et al. [6] through thermoplastic forming. Similar to micro-/nano-scale hot-imprinting, the TPF-based blow molding has also been used to fabricate ultra-smooth and symmetric 3-D metallic glass resonators, which demonstrates precision over 5 orders of magnitude without the use of cleanroom facilities or traditional microfabrication techniques, displaying potential applications in future MEMS vibratory devices, such as accelerometers and gyroscopes, with reduced energy dissipation mechanisms, increased performance and low costs [61].

The thermoplastic micro-forming technique also exhibits great potential in fabrication super-hydrophobic surfaces with long lifespan in service, as demonstrated by Li et al. [62, 63]. Who found that without any modification or post-treatment, superhydrophobic surfaces with good stability could be fabricated by hot-embossing honeycomb patterns on $Pd_{40}Cu_{30}Ni_{10}P_{20}$ MG [62]. By constructing micro-/nano-hierarchical structures on $Zr_{35}Ti_{30}Be_{26.75}Cu_{8.25}$ MG surface, Li et al. [63] not only fabricated superhydrophobic MG surface with water contact angle over 150°, but also found that these surfaces exhibit strong adhesion with water droplets. The combined properties of both superhydrophobicity and strong adhesion toward liquid exhibit promising applications as dry adhesives and transport of liquid micro-droplets, as well as desirable mechanical and corrosion resistance showing potential applications in modern industries [64]. Furthermore, Li et al. revealed that MGs surfaces with hot-embossed textures exhibit low friction coefficient especially under dry contact (see **Figure 8**), which indicates that the lifetime of the textured surfaces could be optimized by minimizing friction [65].

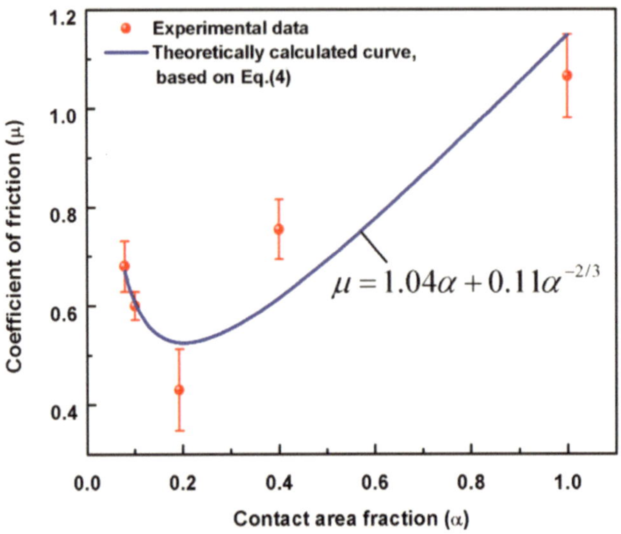

Figure 8. The experimental and theoretically calculated coefficients of friction with the contact area fraction [65].

5. Summary and outlook

Thermoplastic forming provides a promising method to fabricate MGs topological structures and components at various scale sizes, which provides alluring prospects in broadening the application of MGs. The chapter reviews some crucial issues such as the thermoplastic formability, processing techniques and potential applications. Some challenges still exist and impede the practical applications: (1) only few amorphous alloys with excellent glass forming ability, anti-oxidation ability and wide supercooled liquid region, and so on can meet the requirement of thermoplastic forming, (2) the current TPF techniques face challenges in fabricating complicated 3-D structures, (3) the material flow is seriously affected by the interfacial effect on the micro- and nano-scale and the root physical mechanism remains vaguely understood and needs to be settled, and (4) large-scale manufacture is necessary to improve productivity and reduce the cost, if the market of commercial application is developed. Therefore, developing a novel forming technique becomes urgently necessary to breakthrough the alloy systems' limitations. Recent literatures [27–29] have revealed that additive manufacturing (3D printing) is a promising technique for the production of bulk metallic glass (BMG) components without size and alloy system limitations. The authors believe micro-3D printing would provide new opportunities for the creation of small, complex and free-form MG components that were previously unachievable, which would open a new window for MGs fabrication.

Acknowledgements

This work was financially supported by the National Nature Science Foundation of China under Grant nos. 51671090. The authors are also grateful to the Analytical and Testing Center, Huazhong University of Science and Technology for technical assistance.

Conflict of interest

The authors declare that they have no competing financial interests.

Author details

Ning Li[1]* and Jiang Ma[2]

*Address all correspondence to: hslining@mail.hust.edu.cn

1 State Key Laboratory for Materials Processing and Die and Mould Technology, School of Materials Science and Engineering, Huazhong University of Science and Technology, Wuhan, PR China

2 Guangdong Provincial Key Laboratory of Micro/Nano Optomechatronics Engineering, College of Mechatronics and Control Engineering, Shenzhen University, Shenzhen, PR China

References

[1] Li N, Liu Z, Wang XY, Zhang M. Vibration-accelerated activation of flow units in a Pd-based bulk metallic glass. Materials Science and Engineering A. 2017;**692**:62-66. DOI: 10.1016/j.msea.2017.03.062

[2] Guo H, Yan PF, Wang YB, Tan J, Zhang ZF, Sui ML, Ma E. Tensile ductility and necking of metallic glass. Nature Materials. 2007;**6**:735-739. DOI: 10.1038/nmat1984

[3] Jang D, Greer JR. Transition from a strong-yet-brittle to a stronger-and-ductile state by size reduction of metallic glasses. Nature Materials. 2010;**9**:215-219. DOI: 10.1038/nmat2622

[4] Sohn S, Jung Y, Xie Y, Osuji C, Schroers J, Cha JJ. Nanoscale size effects in crystallization of metallic glass nanorods. Nature Communications. 2015;**6**:8167. DOI: 10.1038/ncomms9157

[5] Kumar G, Desai A, Schroers J. Bulk metallic glass: The smaller the better. Advanced Materials. 2011;**23**:461-476. DOI: 10.1002/adma.201002148

[6] Li N, Chen W, Liu L. Thermoplastic micro-forming of bulk metallic glasses: Areview. Journal of Metals. 2016;**68**:1246-1261. DOI: 10.1007/s11837-016-1844-y

[7] Li N, Chen Q, Liu L. Size dependent plasticity of a Zr-based bulk metallic glass during room temperature compression. Journal of Alloys and Compounds. 2010;**493**:142-147. DOI: 10.1016/j.jallcom.2009.12.174

[8] Schroers J. Processing of bulk metallic glass. Advanced Materials. 2010;**22**:1566-1597. DOI: 10.1002/adma.200902776

[9] Kumar G, Tang HX, Schroers J. Nanomoulding with amorphous metals. Nature. 2009; **457**:868-872. DOI: 10.1038/nature07718

[10] He JJ, Li N, Tang N, Wang XY, Zhang C, Liu L. The precision replication of a microchannelmould by hot-embossing a Zr-based bulk metallic glass. Intermetallics. 2012;**21**:50-55. DOI: 10.1016/j.intermet.2011.10.001

[11] Li N, Li DJ, Liu L. Correlation between flow characteristics and interfacial friction behaviour of a Zr-based metallic glass during micro-extrusion. Philosophical Magazine. 2013;**93**:1859-1872. DOI: 10.1080/14786435.2012.762470

[12] Saotome Y, Miwa S, Zhang T, Inoue A. The micro-formability of Zr-based amorphous alloys in the supercooled liquid state and their application to micro-dies. Journal of Materials Processing Technology. 2001;**113**:64-69. DOI: 10.1016/S0924-0136(01)00605-7

[13] Hasan M, Schroers J, Kumar G. Functionalization of metallic glasses through hierarchical patterning. Nano Letters. 2015;**15**:963-968. DOI: 10.1021/nl504694s

[14] Wu X, Li JJ, Zheng ZZ, Liu L, Li Y. Micro-back-extrusion of a bulk metallic glass. Scripta Materialia. 2010;**63**:469-472. DOI: 10.1016/j.scriptamat.2010.05.004

[15] Li N, Li DJ, Wang XY, Liu L. Size-dependent flowing characteristics of a Zr-based bulk metallic glass in the supercooled liquid region. Journal of Alloys and Compounds. 2012;**523**:146-150. DOI: 10.1016/j.jallcom.2012.01.136

[16] Sarac B, Bera S, Balakin S, Stoica M, Calin M, Eckert J. Hierarchical surface patterning of Ni- and Be-free Ti- and Zr-based bulk metallic glasses by thermoplastic net-shaping. Materials Science and Engineering C. 2017;**73**:398-405. DOI: 10.1016/j.msec.2016.12.059

[17] Hasan M, Kumar G. High strain rate thermoplastic demolding of metallic glasses. Scripta Materialia. 2016;**123**:140-143. DOI: 10.1016/j.scriptamat.2016.06.021

[18] Bera S, Sarac B, Balakin S, Ramasamy P, Stoica M, Calin M, Eckert J. Micro-patterning by thermoplastic forming of Ni-free Ti-based bulk metallic glasses. Materials and Design. 2017;**120**:204-211. DOI: 10.1016/j.matdes.2017.01.080

[19] Saotome Y, Iwazaki H. Superplastic backward microextrusion of microparts for micro-electro-mechanical systems. Journal of Materials Processing Technology. 2001;**119**:307-311. DOI: 10.1016/S0924-0136(01)00957-8

[20] Chu JP, Wijaya H, Wu CW, Tsai TR, Wei CS, Nieh TG, Wadsworth J. Nanoimprint of gratings on a bulk metallic glass. Applied Physics Letters. 2007;**90**. DOI: 10.1063/1.2431710

[21] Nishiyama N, Inoue A. Glass transition behavior and viscous flow working of Pd40Cu30Ni10P20 amorphous alloy. Materials Transactions Jim. 1999;**40**:64-71. DOI: 10.1016/S0140-6701(00)96797-3

[22] Kawamura Y, Kato H, Inoue A, Masumoto T. Full strength compacts by extrusion of glassy metal powder at the supercooled liquid state. Applied Physics Letters. 1995; **67**:2008. DOI: 10.1063/1.114769

[23] Bardt JA, Bourne GR, Schmitz TL, Ziegert JC, Sawyer WG. Micromolding three-dimensional amorphous metal structures. Journal of Materials Research. 2007;**22**:339-343. DOI: 10.1557/jmr.2007.0035

[24] Henann DL, Srivastava V, Taylor HK, Hale MR, Hardt DE, Anand L. Metallic glasses: Viable tool materials for the production of surface microstructures in amorphous polymers by micro-hot-embossing. Journal of Micromechanics and Microengineering. 2009;**19**:115030. DOI: 10.1088/0960-1317/19/11/115030

[25] Chen JK, Chen WT, Cheng CC, Yu CC, Chu JP. Metallic glass nanotube arrays: Preparation and surface characterizations. Materials Today. 2018;**21**:178-185. DOI: 10.1016/j.mattod.2017.10.007

[26] Chen W-T, Manivannan K, Yu C-C, Chu JP, Chen J-K. Fabrication of an artificial nanosucker device with a large area nanotube array of metallic glass. Nanoscale. 2018;**10**:1366-1375. DOI: 10.1039/C7NR07360G

[27] Li N, Zhang JJ, Xing W, Ouyang D, Liu L. 3D printing of Fe-based bulk metallic glass composites with combined high strength and fracture toughness. Materials & Design. 2018;**143**:285-296. DOI: 10.1016/j.matdes.2018.01.061

[28] Ouyang D, Li N, Liu L. Structural heterogeneity in 3D printed Zr-based bulk metallic glass by selective laser melting. Journal of Alloys and Compounds. 2018;**740**:603-609. DOI: 10.1016/j.jallcom.2018.01.037

[29] Ouyang D, Li N, Xing W, Zhang JJ, Liu L. 3D printing of crack-free high strength Zr-based bulk metallic glass composite by selective laser melting. Intermetallics. 2017;**90**:128-134. DOI: 10.1016/j.intermet.2017.07.010

[30] Schroers J. The superplastic forming of bulk metallic glasses. Journal of Metals. 2005;**57**:35-39. DOI: 10.1007/s11837-005-0093-2

[31] Schroers J. On the formability of bulk metallic glass in its supercooled liquid state. Acta Materialia. 2008;**56**:471-478. DOI: 10.1016/j.actamat.2007.10.008

[32] Bryn Pitt E, Kumar G, Schroers J. Temperature dependence of the thermoplastic formability in bulk metallic glasses. Journal of Applied Physics. 2011;**110**:043518. DOI: 10.1063/1.3624666

[33] Lu J, Ravichandran G, Johnson WL. Deformation behavior of the Zr41.2-Ti13.8-Cu12.5-Ni10-Be-22.5 bulk metallic glass over a wide range of strain-rates and temperatures. Acta Materialia. 2003;**51**:3429-3443. DOI: 10.1016/S1359-6454(03)00164-2

[34] Li N, Chen Y, Jiang MQ, Li DJ, He JJ, Wu Y, Liu L. A thermoplastic forming map of a Zr-based bulk metallic glass. Acta Materialia. 2013;**61**:1921-1931. DOI: 10.1016/j.actamat.2012.12.013

[35] Saotome Y, Itoh K, Zhang T, Inoue A. Superplastic nanoforming of Pd-based amorphous alloy. Scripta Materialia. 2001;**44**:1541-1545. DOI: 10.1016/S1359-6462(01)00837-5

[36] Saotome Y, Imai K, Shioda S, Shimizu S, Zhang T, Inoue A. The micro-nanoformability of Pt-based metallic glass and the nanoforming of three-dimensional structures. Intermetallics. 2002;**11**:1241-1247. DOI: 10.1016/S0966-9795(02)00135-8

[37] Kato H, Wada T, Hasegawa M, Saida J, Inoue A, Chen HS. Fragility and thermal stability of Pt- and Pd-based bulk glass forming liquids and their correlation with deformability. Scripta Materialia. 2006;**54**:2023-2027. DOI: 10.1016/j.scriptamat.2006.03.025

[38] Böhmer R, Ngai KL, Angell CA, Plazek DJ. Nonexponential relaxations in strong and fragile glass formers. The Journal of Chemical Physics. 1993;**99**:4201-4209. DOI: 10.1063/1.466117

[39] Wang XL, Li N, Dai WL, Zhang M, Gong P. Thermoplastic micro-formability of TiZrHfNiCuBe high entropy metallic glass. Journal of Materials Science & Technology. 2018;**25**:4-11. DOI: 10.1016/j.jmst.2018.04.006

[40] Lu J, Ravichandran G, Johnson WL. Metallic glass over a wide range of strain-rates and temperatures. 2003;**51**:3429-3443. DOI: 10.1016/S1359-6454(03)00164-2

[41] Kawamura Y, Nakamura T, Inoue A. Superplasticity in Pd40Ni40P20 metallic glass. Scripta Materialia. 1998;**39**:301-306. DOI: 10.1016/S1359-6462(98)00163-8

[42] Wiest A, Harmon JS, Demetriou MD, Dale Conner R, Johnson WL. Injection molding metallic glass. Scripta Materialia. 2009;**60**:160-163. DOI: 10.1016/j.scriptamat.2008.09.021

[43] Martinez R, Kumar G, Schroers J. Hot rolling of bulk metallic glass in its supercooled liquid region. Scripta Materialia. 2008;**59**:187-190. DOI: 10.1016/j.scriptamat.2008.03.008

[44] Schroers J, Pham Q, Peker A, Paton N, Curtis RV. Blow molding of bulk metallic glass. Scripta Materialia. 2007;**57**:341-344. DOI: 10.1016/j.scriptamat.2007.04.033

[45] Schroers J, Hodges TM, Kumar G, Raman H, Barnes AJ, Pham Q, Waniuk TA. Thermoplastic blow molding of metals. Materials Today. 2011;**14**:14-19. DOI: 10.1016/S1369-7021(11)70018-9

[46] http://www.schroerslab.com/research/processing-of-bmgs/another-sub-item-of-processing-bmgs

[47] Johnson WL, Kaltenboeck G, Demetriou MD, Schramm JP, Liu X, Samwer K, Kim CP, Hofmann DC. Beating crystallization in glass-forming metals by millisecond heating and processing. Science. 2011;**332**:828-833. DOI: 10.1126/science.1201362

[48] Kaltenboeck G, Demetriou MD, Roberts S, Johnson WL. Shaping metallic glasses by electromagnetic pulsing. Nature Communications. 2016;**7**:10576. DOI: 10.1038/ncomms-10576

[49] Hung JC, Huang CC. Evaluation of friction in ultrasonic vibration-assisted press forging using double cup extrusion tests. International Journal of Precision Engineering and Manufacturing. 2012;**13**:2103-2108. DOI: 10.1007/s12541-012-0278-x

[50] Li N, Xu XN, Zheng ZZ, Liu L. Enhanced formability of a Zr-based bulk metallic glass in a supercooledliquid state by vibrational loading. Acta Materialia. 2014;**65**:400-411. DOI: 10.1016/j.actamat.2013.11.009

[51] Ma J, Liang X, Wu X, Liu Z, Gong F. Sub-second thermoplastic forming of bulk metallic glasses by ultrasonic beating. Scientific Reports. 2015;**5**:17844. DOI: 10.1038/srep17844

[52] Carmo M, Sekol RC, Ding S, Kumar G, Schroers J, Taylor AD. Bulk metallic glass nanowire architecture for electrochemical applications. ACS Nano. 2011;**5**:2979-2983. DOI: 10.1021/nn200033c

[53] Sekol RC, Kumar G, Carmo M, Gittleson F, Hardesty-Dyck N, Mukherjee S, Schroers J, Taylor AD. Bulk metallic glass micro fuel cell. Small. 2013;**9**:2081-2085. DOI: 10.1002/smll.201201647

[54] Huang JC, Chu JP, Jang JSC. Recent progress in metallic glasses in Taiwan. Intermetallics. 2009;**17**:973-987. DOI: 10.1016/j.intermet.2009.05.004

[55] Ma J, Yi J, Zhao DQ, Pan MX, Wang WH. Large size metallic glass gratings by embossing. Journal of Applied Physics. 2012;**112**. DOI: 10.1063/1.4752399

[56] Inoue A, Takeuchi A. Recent development and application products of bulk glassy alloys. Acta Materialia. 2011;**59**:2243-2267. DOI: 10.1016/j.actamat.2010.11.027

[57] Pan CT, Wu TT, Chang YC, Huang JC. Experiment and simulation of hot embossing of a bulk metallic glass with low pressure and temperature. Journal of Micromechanics and Microengineering. 2008;**18**:25010. DOI: 10.1088/0960-1317/18/2/025010

[58] Bardt J, Mauntler N, Bourne G, Schmitz TL, Ziegert JC, Sawyer WG. Metallic glass surface patterning by micro-molding. American Society of Mechanical Engineers, Manufacturing Engineering Division, MED. 2005:1123-1129. DOI:10.1115/IMECE2005-81099

[59] Liu X, Shao Y, Lu SY, Yao KF. High-accuracy bulk metallic glass mold insert for hot embossing of complex polymer optical devices. Journal of Polymer Science, Part B: Polymer Physics. 2015;**53**:463-467. DOI: 10.1002/polb.23670

[60] Ma J, Zhang X, Wang WH. Metallic glass mold insert for hot embossing of polymers. Journal of Applied Physics. 2012;**112**:024506. DOI: 10.1063/1.4737484

[61] Kanik M, Bordeenithikasem P, Kim D, Selden N, Desai A, M'Closkey R, Schroers J. Metallic glass hemispherical shell resonators. Journal of Microelectromechanical Systems. 2015;**24**:19-28. DOI: 10.1109/JMEMS.2014.2363581

[62] Xia T, Li N, Wu Y, Liu L. Patterned superhydrophobic surface based on Pd-based metallic glass. Applied Physics Letters. 2012;**101**:81601. DOI: 10.1063/1.4747327

[63] Li N, Xia T, Heng L, Liu L. SuperhydrophobicZr-based metallic glass surface with high adhesive force. Applied Physics Letters. 2013;**102**:251603. DOI: 10.1063/1.4812480

[64] Ma J, Zhang XY, Wang DP, Zhao DQ, Ding DW, Liu K, Wang WH. Superhydrophobic metallic glass surface with superior mechanical stability and corrosion resistance. Applied Physics Letters. 2014;**104**:173701. DOI: 10.1063/1.4874275

[65] Li N, Xu E, Liu Z, Wang X, Liu L. Tuning apparent friction coefficient by controlled patterning bulk metallic glasses surfaces. Scientific Reports. 2016;**6**:1-9. DOI: 10.1038/srep39388

Section 3

Advanced Simulation

Chapter 3

Mechanical Properties and Deformation Behaviors of Metallic Glasses Investigated by Atomic-Level Simulations

Hongwei Zhao, Dan Zhao, Bo Zhu and
Shunbo Wang

Additional information is available at the end of the chapter

http://dx.doi.org/10.5772/intechopen.76830

Abstract

The chapter reviewed recent developments about intrinsic structure of metallic glasses and their mechanical properties at atomic level, with an emphasis on making connections between developments in theory and simulations. Topics covered the following: structure analysis on metallic glasses with methods of pair distribution function, Honeycutt-Andersen analysis, and Voronoi tessellation; the connection of structure with the mechanical properties; shear band initiation and the development at the atomic level; and deformation mode transition from cast metallic glasses to reconstructed nanoglasses. These works provided theoretical understanding on the essence of metallic glasses' mechanical properties and deformation behaviors, and offered promises for more extensive applications of metallic glasses.

Keywords: metallic glasses, molecular dynamics simulation, shear bands, structure analysis, mechanical behavior

1. Introduction

Metallic glasses (MGs), also named amorphous alloys for their disordered atomic structure, behave excellent mechanical behaviors, including high elastic modulus, high yield stress, and extreme high hardness, compared with traditional crystalline alloys. Considering their excellent mechanical performance in micro-nano scale, MGs have a great application potential in ultra-precision systems. The research work on MGs has been proceeding for over 50 years both experimentally and theoretically. Meanwhile, the corresponding simulation methods with aids of computer science have also been developed to provide assistance for the experimental research from the theoretical aspect. The most used simulation methods for the MGs are finite element method (FEM) and molecular dynamics (MD) method. FEM constructs a constitutive model of MG at the macroscopic scale on time and space, which is coincident with experimental works. For this advantage, it is usually employed to investigate the mechanical properties and shear bands (SBs) formation in the bulk MG systems [1–4]. Although a MG is usually regarded as isotropic and homogeneous from the macroscopic scale, a homogeneous constitutive model is not sufficient for the generation and localization of SBs in a typical brittle MG system. For the improvement, two different SB formation theories, free volume theory [1, 2] and shear transition zone (STZ) theory [3, 4] were employed in constitutive model to provide nucleation possibilities in the MG model, and they are proved quiet effective when depicting the formation process of SBs. However, these FEM simulations are basing on a fuzzy description of flow events according to the two immature theories, but neglect intrinsic dynamics of the flow events from atomic aspect. Furthermore, the theoretical frames and constitutive model for the research of MGs have to build on accurate description of their atomic structure. Since, the lack of effective experimental tools on characterizing the atomic structure of MGs, resolving the intrinsic structure only from experimental methods is unreasonable. Recently, MD methods are developed to offer possibility for the investigation on MGs from atomic level. By simulating the atoms movements and tracking atoms trajectories, MD simulation can calculate the basic thermal information of a MG ensemble, including temperature, potential energy, as well as some derived information such as atom shear stress, atom shear strain, and structure evolution and so on. Thus, MD simulation method is an effective tool when investigating the intrinsic structure of MG, and its connection with mechanical properties, as well as the elastic to plastic deformation transition.

In this chapter, we mainly take a review on the former research work on the connections between structure and properties in MGs. We began with the current understanding in the structure of MGs and then discuss their proper connection with some mechanical behaviors. Afterwards, more specific discussion on the SBs formation and development will be given based on various mechanics of MD simulations. Finally, based on the limitation of MGs in their applications, we discussed several multi-component materials derived from cast MGs, and have an outlook on the development trend of MG preparation and application.

2. Structure of metallic glasses

The essential feature of MG is considered as disordered, but there is still no exactly model to describe the total features of amorphous structure. The little research on the atomic structure of amorphous system, as well as the correlation of amorphous structure and mechanical behaviors, has obstructed the development on design and implication of amorphous materials seriously. Thus, the characterization and modeling of amorphous structure has been one of the most important and challenge research works in condense physics and material science area, and it is also the basis of research work on the mechanical behavior of MGs.

The first model of atomic structure in MG was developed by Bernal, and it was described as a dense random packed structure of equal-sized spheres with the absence of medium-range and long-range order [5]. Bernal used the active hard balls to describe a liquid-like amorphous system, and analyzed the radial distribution function of the model, and found to have a good consistence with that of liquid system. This model provided a possibility of modeling and characterization based on the computer science. The further development of amorphous structure modeling has been based on Bernal's work, and more developed characterization methods have been raised from the aspects of statistic and topology with the aids of MD simulations. The most used characterization methods of MGs with MD simulation are presented as the following in this section, including pair distribution function, Honeycutt-Andersen analysis, and Voronoi tessellation method.

2.1. Pair distribution function

Pair distribution function (PDF) is the most classical and important statistical analysis method in amorphous systems; it is widely used in the characterization of liquids and amorphous materials. It is a pair correlation representing the probability of finding atoms, and described as a function of distance r from an average center atom. In a monatomic system, PDF is defined as:

$$g(r) = \frac{1}{4\pi r^2 \rho N} \sum_{i=1}^{N} \sum_{j=1, j\neq i}^{N} \delta\left(r - \left|\bar{r}_{ij}\right|\right), \quad (1)$$

where $\left|\bar{r}_{ij}\right|$ is the interatomic distance between atom i and atom j, and ρ is the number density of atoms in the system with N atoms. According to the PDF curve of a MG, the short-to-medium range order information can be manifested by the peak position, peak width, and relative intensity, etc. Conventionally, the nearest-neighbor shell atoms contribute to the first peak, which represents short range order (SRO) in the MG. A further distance up to 1–2 nm contributes the medium-range order (MRO). With distance r going larger, PDF gradually converges to unity, which means the atoms are randomly distributed, represents the long-range disorder. The structure evolution from liquid to glass can be shown from the PDF curves during the cooling procedure. **Figure 1** displays the PDF curves of a binary $Cu_{64}Zr_{36}$ MG in different temperature regions during its preparation process. It can be clearly seen that with decreasing temperature during quenching process, the second peak of total PDF gradually splits and becomes more pronounced, which indicates the formation of the glass phase and enhancement of SRO. Furthermore, the appearing temperature of the split is usually relevant to the glass transition temperature (T_g), indicating the generally formation and stability of the glass structure.

PDF is concise and clear for the characterization of amorphous materials, especially when detecting the amorphous phase transferred from a crystalline system; thus, PDF method is widely used to verify the effective MG models in MD simulations [6]. However, this method is only established on a statistic basis, without considering the exactly certain structure of MGs. It cannot provide the specific topology description of a certain amorphous system, thus has a limitation when further revealing the atomic structure geometry of MGs.

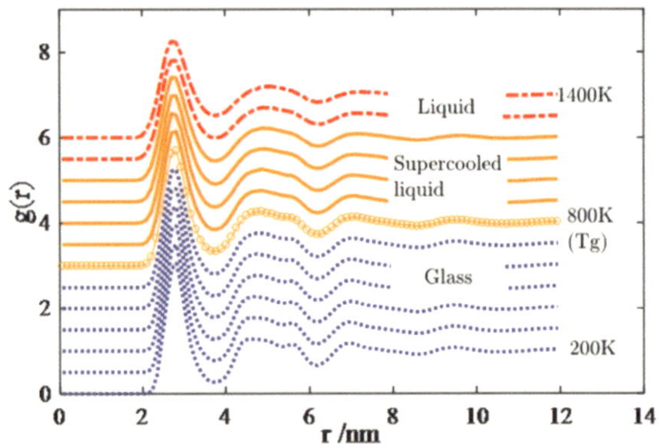

Figure 1. Partial distribution function of Cu-Zr MG during quenching process.

2.2. Honeycutt-Andersen analysis

A widely used and effective analysis method for the topology structure of crystalline systems is common neighbor analysis (CNA), which is proposed by Clarke and Jónsson [7]. CNA is mainly to describe a bonding correlation between neighboring atoms in highly regular ordered systems with integers *jkl*: (a) *j* for the number of near neighbors they have in common; (b) *k* for the number of bonds among the shared neighbors; (c) and *l* represents the number of bonds in the longest bonding chain in *k* type bonds. It can effectively detect general structure and abnormal structure in crystalline materials, but when used for various disordered structures, it cannot describe an accurate amorphous local structure and might have a misleading to structure analysis of MG.

Similar with CNA analysis method, Honeycutt-Andersen (H-A) analysis method uses a multiple integers *ijkl* to describe more unregularly ordered and complicated bond types [8]. The additional number *i* in the H-A indices *ijkl* indicates whether or not the near neighbors are bonded, $i = 1$ for bonded pairs and $i = 2$ for nonbonding atoms, and *l* is used to distinguish different bond geometries in case the first three numbers are same. The cutoff distance is derived from PDF curves for the particular pair of atoms. Different structures have different H-A indices to represent distinguished bond types. In the MGs, 1551 and 2331 are usually the characteristics of icosahedral ordering, which mainly contribute to SRO in MGs.

H-A indices can be employed to analyze SRO structures of MG. Duan et al. took the H-A analysis on $Cu_{50}Zr_{50}$ MG during the preparation process [9]. At the liquid state, the bond types with H-A indices 1441 and 1661 dominated the initial B2 phase crystalline model, and had the proportions 42 and 56%, respectively. When the temperature went to the totally melting liquid phase, the proportions dropped to 7 and 6%. The amorphous bond type 1431, 1541, and 1551 increased to 16, 12, and 14%, respectively. During a quenching process, the amorphous H-A indices were monitored, they found out that 1431 and 1541 pairs do not change much, while the icosahedral 1551 and 2331 pairs increased uniformly as the system being supercooled until a local maxima at 700 K, indicating the final state came to be stable amorphous state. Therefore, as the MG system is cooled from its liquid state, the icosahedral symmetry keeps increasing and this SRO structure has been proved a correlation with stability and generation of a MG system.

2.3. Voronoi tessellation

H-A analysis method can reveal the topology SRO structure effectively, but fail to obtain the specific geometry features of SRO structure. Voronoi tessellation is a polyhedral analysis method developed by introducing the concept of Voronoi polyhedral, which is defined as a closed convex polyhedral enclosed by the vertical bisectors of the nearest-neighbor atoms and center atom [10]. Usually, i_k is used to represent the number of polygon with k edges in a Voronoi polyhedral, and a four-number vector $< i_3, i_4, i_5, i_6 >$ is chosen as the Voronoi index, to describe the arrangement and symmetry of the nearest-neighbor atoms around the center atom. For example, an icosahedral with Voronoi index < 0,0,12,0 > has 12 pentagons enclosed, and a polyhedral with index < 0,3,6,0 > has 3 quadrilaterals and 6 pentagons enclosed. The polyhedral with same Voronoi index can be concluded as one type cluster. After analyzing all the local clusters compositions in MG, the motifs are summarized as a series of typical Voronoi polyhedral types and can be regard as mainly contributions to SRO structure.

Among these mentioned analysis methods, the Voronoi tessellation method is the most accurate and effective tool for typical structure identification in MGs, and also be used by many researchers [10–13]. Cheng et al. used this method to describe the structure evolution of Cu-Zr MG during quenching process [11]. Five most popular Voronoi polyhedral types were elected, which accounting for over 75% of all Cu-centered polyhedral. These clusters had common characteristics of high symmetry and a CN of 12, while the other fragmented clusters have low symmetry and usually irregular polyhedral with unfavorable CNs or highly distorted shapes. The polyhedral with Voronoi index < 0,0,12,0 > have a symmetrical fivefold geometry and icosahedral structure, and defined as "full icosahedra" (FI). FI clusters were found increase sharply during the quenching process, especially during the supercooled liquid regime. While the cluster motifs of MG with lower symmetry had a gradually change with the sample cooling. But the fragmental ones, which might the motifs of equilibrium liquid at melting temperature, decreased sharply accompanying with formation of glass state. These observations suggested that FI cluster mostly contributed to SRO and dominated the solid-like nature of MGs, and with lower cooling rate, the fraction of FI increased and made the generated MG a more stable one, which was consistent with glass formation cooling rate dependence. We will have a more detailed discuss on the cooling rate effect in Section 3.

2.4. MRO structures in MG

By combining experimental characterization tools such as XRD, EXAFS, and RMC together with *ab initio* simulation method, Sheng et al. resolved the MRO structure of various MGs [6]. With application of Voronoi tessellation, the formation of local solute-centered coordination polyhedral with strong chemical favoring unlike bonds was found and defined with "quasi-equivalent cluster". On the basis of the quasi-equivalent clusters, the MRO was further proposed to be dense packing of these clusters by sharing shell atoms, and regard as general feature of MG. Three basic icosahedra linkages were concluded as face-sharing (FS), edge-sharing (ES) and vertex-sharing (VS). **Figure 2** shows the typical connections of SRO and MRO structure in three MGs. Liu analyzed various MGs structure by PDF data and MD simulation. They indicated that the MRO structure can be interpreted as spherical-periodic

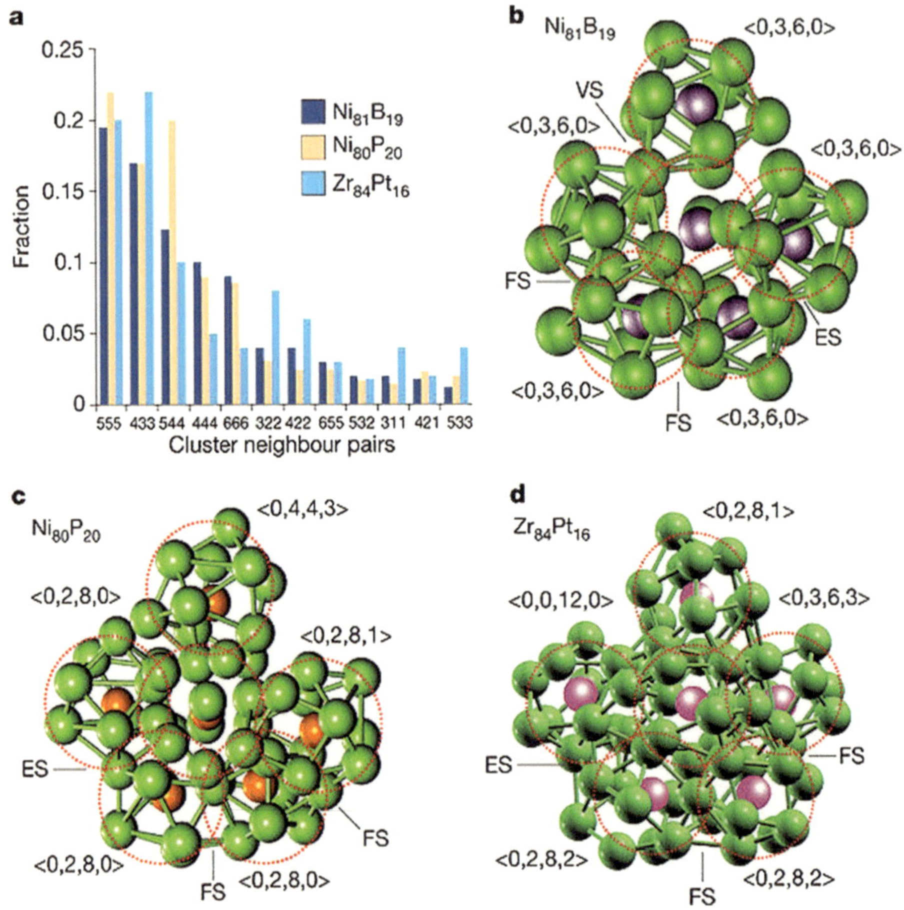

Figure 2. Typical connections of SRO clusters and MRO structures in MGs [6].

order and local translational symmetry, and the increasing of MRO structures proceeded throughout the glass transition process [14].

The combining of PDF and topological analysis method can also reveal the formation of MRO. Liang identified icosahedral clusters from an Mg-Zn MG using H-A analysis method, and found some neighbor clusters have linkage with structure packed and atoms shared [15]. The icosahedra linkages of intercross-sharing (IS), FS, ES, and VS were also detected. By evaluating the average distances between center atoms in each linking cluster, Liang found the distribution of the distances had a coincidence with the second peak splitting of PDF curve. With more clusters interacted into a network, MRO came to formation in the glass.

With the increase in simulation dimension, it was also found that the clusters tend to form large and interconnecting networks. Ward further extended the range of correlated packing of the icosahedral clusters, and found the diversities of the network diffusivity in compositions about the binary Cu-Zr MGs [12]. With 30% fraction of Zr atoms, the MG had higher fraction of Cu-centered FI clusters, which tended to form a single large network. With the promotion of

Zr atoms, the fraction of Cu-centered FI decreased severely, and the formed networks tended to be diffusive with decrease in size and increase in number. At 50% fraction for Zr atoms, the number of networks reached the maximum and had a downward tendency with the further increase of Zr atoms fraction. It was noted that the compositions of $Cu_{70}Zr_{30}$ and $Cu_{50}Zr_{50}$ both had high glass-forming ability, differences in network forming proved that the presence of networks is neither a sufficient nor necessary condition for the formation of glass phase, but networks formation does have dependence on the preparing procedures, and a strong effect on mechanical behavior of MG as well. The detailed explanation will be given in Section 3.

Another found by Cheng is that an addition of a small percentage of Al in the original Cu-Zr binary MG can lead to dramatically increased population of FI clusters and their spatial connectivity (shown in **Figure 3**) [12]. It has been known that the ideal icosahedral dense packing requires several factors: an atomic size ratio of 0.902 for a hard-packing model [16], and negative mixing heat of atoms. For binary Cu-Zr MG, the atomic size ratio is 0.804, Al atom has

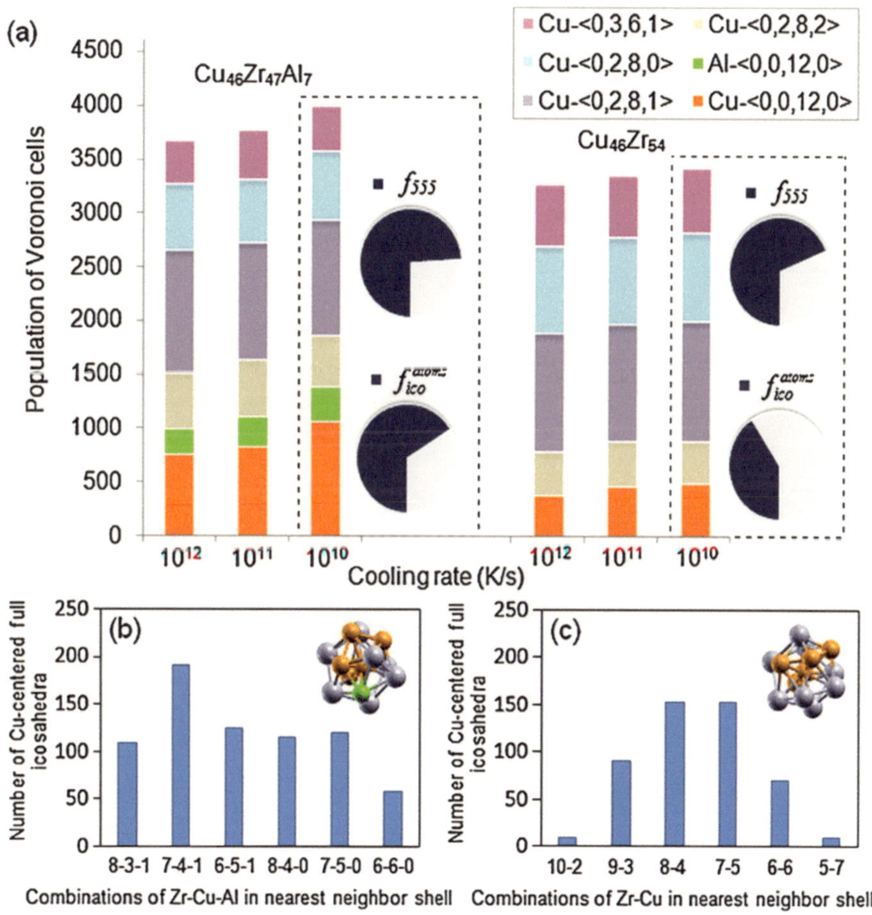

Figure 3. (a) Histogram showing differences in SRO components and fractions in Cu-Zr-Al ternary MG and Cu-Zr binary MG; (b) combinations of Cu-Zr-Al atoms in Cu-centered FI SRO in Cu-Zr-Al ternary MG and (c) Cu-Zr binary MG [13].

a radius in between of Zr and Cu atoms, the three atomic sizes can adjust the coordination polyhedron around the Cu atom, and increases the possibility of comfortable arrangements to reach FI. Furthermore, the negative mixing heat of Al with Zr and Cu drives Al to scatter in the Cu-Zr matrix. A more ideal ratio of 0.905 of Zr-Al atoms make Al surrounded by Zr become the topologically optimal way for FI packing. Similar with the binary Cu-Zr MG mentioned above, the FIs in this ternary MG overlapped and connected with the bonding forms of tetrahedral sharing (TS, same as IS), FS, ES, and VS, and then interconnected with each other resulting in the formation of networks (shown in **Figure 4**). The degree of connectivity of FI might serve as the backbone of the MG structures, and induced the more stable structure and improvement of mechanical properties in ternary glass compared with the binary glass. A further study taken by Tang suggested that those SRO clusters with low degree five-fold symmetry structure could also tend to form interconnected networks [17]. But different with the solid-like networks formed by FI clusters, the low five-fold symmetry clusters built up liquid-like networks. Compared with the solid-like networks, the liquid-like ones were less resistant to shear events and usually fertile sites for plastic deformation.

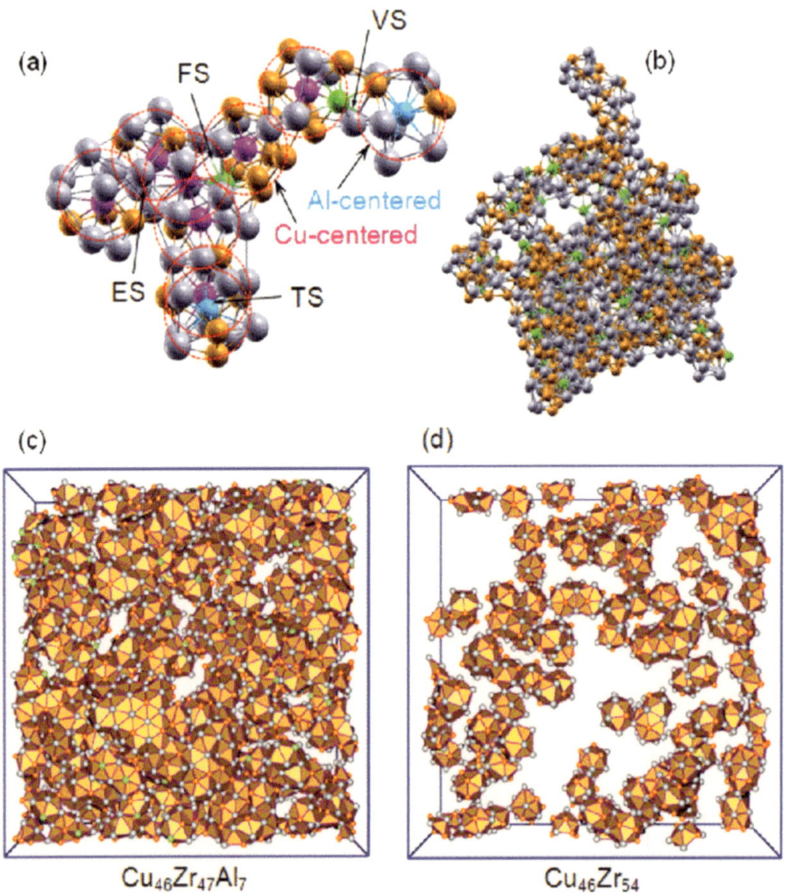

Figure 4. (a) A supercluster consisting four types of connections VS, ES, FS and TS; (b) a large supercluster consisting of over 700 atoms; (c) higher degree of connectivity of FI shown in Cu-Zr-Al MG compared with (d) Cu-Zr MG [13].

http://dx.doi.org/10.5772/intechopen.76830

3. Correlation of structure and mechanical properties

The correlation of structure and mechanical properties is a central theme of materials research, no matter crystalline materials or amorphous ones. The properties of MGs change pronouncedly with the internal structures change. Several works have been done for the influence of changes in composition proportions and variation in the processing conditions for MGs of a fixed composition, but what really governs the properties of MG from the sight of atomic-level structures has to be concerned for its quantitative and predictive theory establishment. The icosahedral cluster structures usually have a more dense packing efficiency, and the distributions are not even in space with connection of icosahedral existing. Thus, the MG structure has intrinsic fluctuations, and the local atomic-level area has heterogeneity in structure and dynamics, and the nanoscale mechanical heterogeneity had been proved with the using of dynamics force microscopy [18]. The mechanical response of MG sample can be significantly influenced by various local structure motifs, relative fractions, and their distributions in space. In other words, the structural heterogeneity necessarily leads to mechanical heterogeneity and decide the mechanical behavior of the MG [10, 19].

According to Wakeda's investigations on the shear deformation of $Cu_xZr_{1\text{-}x}$ (x = 0.30–0.85) MGs, the local geometrical structures had large variation with the change of proportions, and affected the yield and fracture behavior of the MGs [20]. With application of Voronoi polyhedral analysis method, the local structures were characterized by pentagons and free volume, and these two factors were found related to each other. The pentagonal regions corresponded to the densely packed structure and nonpentagonal regions corresponded to the free volume structure. With shear simulation taken on the various MGs, it was found the pentagon-rich region tends to undergo elastic deformation, while the pentagon-poor was easily deformed plastically. Pentagon-rich local area had more formation of fivefold symmetry structure, and more relevant to the FI cluster, thus the aggregation of pentagons could be regard as formation of SRO. The results indicated the pentagonal SRO contributed the structural stability as well as elastic strength, while the opposite ones closely related to yield and fracture behavior in MGs.

The preparation processing of MG also has a significant influence on the formation of SRO cluster types and contents, and induces local SRO structure heterogeneity in MG. Cheng found that the mechanical behavior and dynamics responses varied with different cooling rates of MG [11]. When applied the same tensile loading procedure to the MG samples prepared under different cooling rates, the sample with lower cooling rate showed a strong tendency to form a single and highly localized SB, while the one with higher cooling rate had a more homogeneous deformation. To obtain a quantitative evaluation of strain localization, the degree of strain localization parameter was defined as:

$$\psi = \sqrt{\frac{1}{N}\sum_{i=1}^{N}\left(\eta_i^{Mises} - \eta_{ave}^{Mises}\right)^2}, \tag{2}$$

where, a larger ψ means larger fluctuations in the atomic strain and a more localized deformation mode. The strain localization degree ψ was found negatively correlated with cooling rates and positively correlated with the fraction of Cu-centered FI clusters. A shear deformation simulation also showed the MG with lower cooling rate had higher shear stress compared with higher cooling rates. And the analysis on shear modulus showed a consistent tendency with strain localization phenomenon. These results showed the dynamics heterogeneity have a significant dependence on cooling rate, and revealed that the MG structure and mechanical behavior can be regulated by controlling the quenching process and cooling rates.

Since SBs formation is a significant signal for catastrophic failure of MG, the structure evolution analysis of SBs is necessary for the dynamics investigation during a deformation. Cao et al. studied the correlation of cluster evolution with the shear localization initiation in a Cu-Zr MG in the SB regime [21]. Fractions of several dominant cluster types in the SB forming region were analyzed during deformation, as shown in **Figure 5**. The breakdown of FI clusters was identified as a structural signature of the initiation of shear localization. With distorting of the FI backbone into less-shear-resistant, uncomfortable clusters, the shear localization propagating to a major SB with a velocity close to speed of voice. Feng et al. took quantitative treatment on the atomic structure of SBs in Cu-Zr MG and found the "liquid-like" features of SB [22]. With the method of H-A analysis, bonded pair types and distributions in SB are more similar to supercooled liquid compared with the MG matrix (shown in **Figure 6**). The SRO analysis results showed heterogeneous structure distribution in clusters. Zr-centered <0,2,8,5 > clusters exhibited strong spatial correlations and tendency to connect with each other in the SB, and formed interpenetrating solid-like backbone in SB (shown in **Figure 7**). Ju et al. took another view on the correlation of the mechanical properties and the local structural rearrangement of Mg-Zn-Ca MG by using H-A analysis method [23]. From the H-A index analysis on the system, the numbers of 1551 and 1431 local structures decreased with the increasing strain until 0.05, and the H-A indices of different types remained almost constant when the strain became larger than 0.05. The distributions of atomic local shear strain during the tension process showed the stages of the initialization of STZs, the extension of STZs, and the formation of shear bands along a direction 45° from the tensile direction.

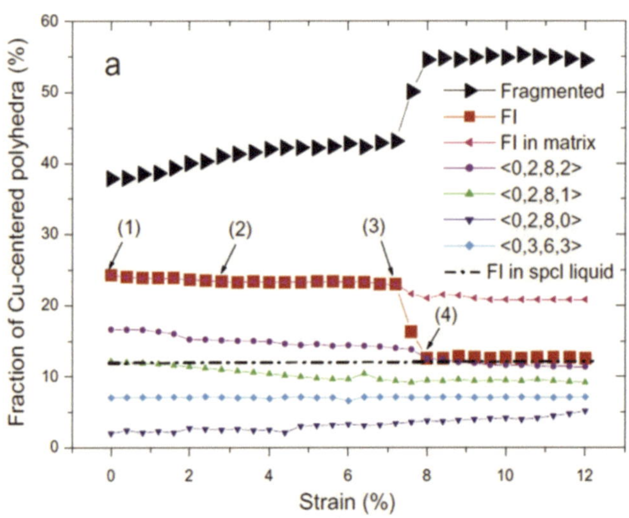

Figure 5. Variations of five dominant SRO clusters with the sample strain increasing in the SB region [21].

http://dx.doi.org/10.5772/intechopen.76830

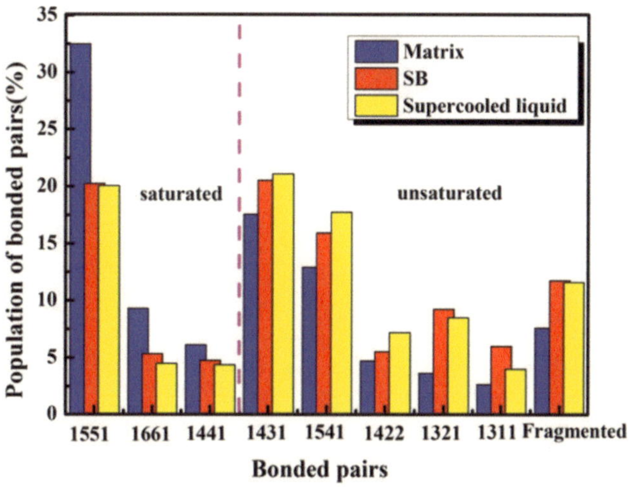

Figure 6. Comparison of several bonded pair types in the matrix, the SB, and supercooled liquid [22].

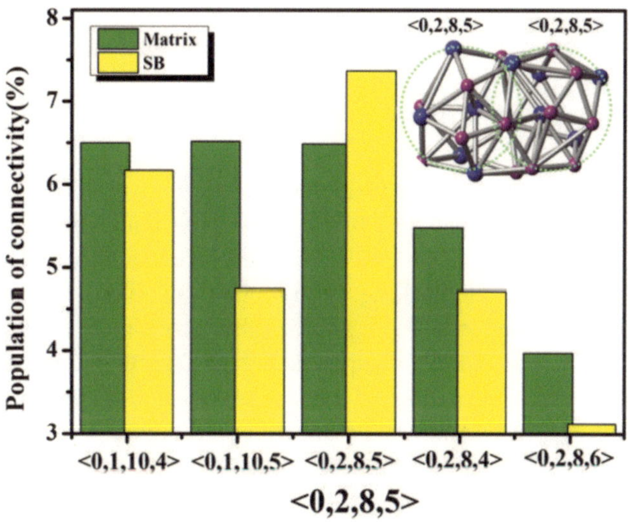

Figure 7. Connectivity of the network formed by the central atoms of the <0,2,8,5 > clusters [22].

The researching work about MG structure revealed the interconnection of icosahedra clusters was the intrinsic nature of MGs. The networks forming by neighboring icosahedra to medium range is negligible for the correlation of structure and mechanical properties in MGs. Lee et al. extended the structural hierarchy to a longer range by connecting MRO structures to large networks [24]. As we mentioned above, the MRO structures are constructed by connecting the neighboring SROs with each other. In Lee's works, interpenetrating connection of icosahedra (ICOI) structure was constructed by icosahedra, and it was illustrated as a typical MRO structure to explore correlation of structure and mechanical properties. To find out which pattern motif mostly contributed to structure stability, the potential energy of the center atom of the individual

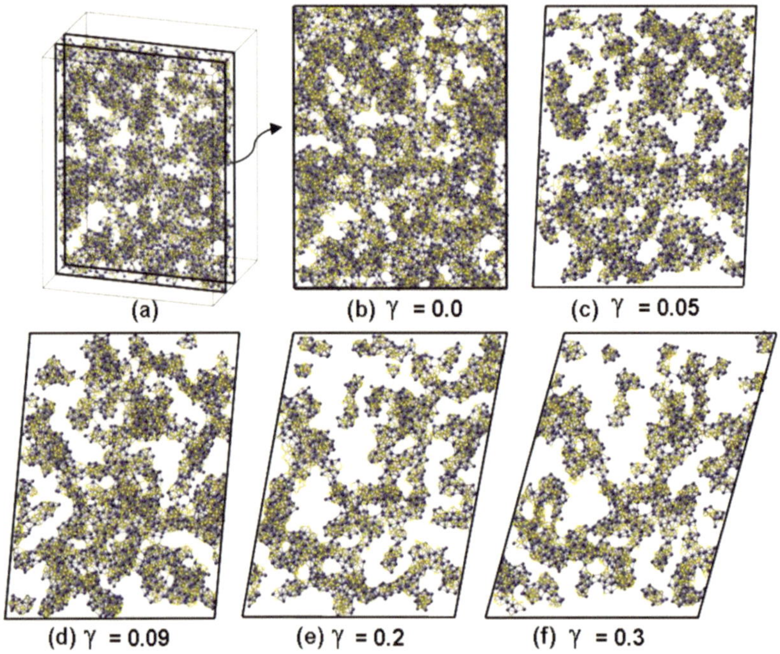

Figure 8. The breaking of ICOI networks in the during shear deformation [24].

icosahedra participating in the connection scheme of each patterns were calculated. It was found the icosahedra with volume shared type had a lower potential energy and stable structure, $Cu_{65}Zr_{35}$ had larger fractions of ICOIs with close connection of ICOIs compared to $Cu_{50}Zr_{50}$, and together with more dense atomic packing. With addition of shear simulations, $Cu_{65}Zr_{35}$ was found more resistant to applied load and showed higher elastic modulus and yield stress. The highly connected MROs constituted a compact icosahedral network over an extended range, which was resistant to stress-induced shear transformations under applied load. The breaking of ICOI networks during shear deformation are shown in **Figure 8**. This work presented the connectivity of MROs and showed its significant influence on the mechanical properties of MG.

4. Shear bands in MGs

This part moves on to a view of larger scale on the shear bands and their connection with the deformation of MGs, which is believed as a breakthrough of atom scale deformation in simulations with experimental phenomenon in macro scale. MGs have a similar performance in experiments with other amorphous materials such as silica glasses and polymers [25, 26], with blunt tip applied on the surface, typical slip line patterns will be shown in the cross surface, and the slip lines are regard caused by shear stress, thus called shear bands.

The formation and spreading of highly localized SBs is widely believed responsible for the failure of brittle MGs. But the ductile MGs could show homogeneous deformation without localization of SBs under the same mechanical loading. To describe the plastic flow and reveal the deformation transition mechanism, the initiation of SBs from a shear transformation is a fundamental problem for the starting. One of the most popular explanations is STZ theory raised by Argon [27]. According to the theory, STZs are the clusters composed of tens to hundreds of atoms, and the basic elements undergoing the plastic flow in MG. When a shear stress is applied on the MG, the atoms in the cluster have relative movements, and deformation takes place inside the cluster. The distortion of STZ is an activation process, and needs a certain activation energy and volume, to transfer a STZ from a high energy state to a lower one.

The propagation of a SB from a collection of STZs is next problem to be concerned. Ogata et al. performed shear deformation on Cu-Zr bulk MG with MD simulations [28], and observed the nucleation and localization of SBs during shear deformation. The nucleation of SB was taken place from a local atomic rearrangement like STZ, and this local deformation induced releasing of elastic strain energy in the shear plane, and activated the generation of more STZs. Then, the STZs interacted with each other and formed a SB in the shear direction. Furthermore, the generation of SB enhanced the elastic strain energy of the surrounding materials in the in-plane directions, made the spreading of SB more localized.

Shimizu et al. described the propagating and developing of SB as a growth process of embryonic shear band (ESB) [29, 30]. An ESB appeared at the concentrator of a group of activated STZs. When the far-field shear stress exceeds the glue traction with temperature rising induced by frictional heating, and the length of the ESB increased to a critical length over 100 nm, the ESB would become maturing to a localized SB. An aged-rejuvenation-glue-liquid (ARGL) SB model was used for the description of the shear front, and four zones were defined as aged glass, rejuvenated glass, glue, and liquid (illustrated in **Figure 9**). The temperature distributes from room temperature at aged glass end to over the glass transition temperature at the liquid end. Their model was supported by the temperature rising phenomenon, which can reach a maximum over 1000 K, can also be verified from the research work by Lewandowski et al. [31]. They planted a fusible tin coating to observe the temperature rise during bend-test on MG, and observed melting in the SBs formation area, thus deduced that local temperature rise in the bands can reach a few thousand kelvin.

For a further understanding on SBs, the mechanical properties of SBs were investigated by Zhou et al. by testing samples with pre-existing SBs with tensile MD simulations [32]. It was demonstrated that pre-deformation lowered material strength and triggered enhanced strain fluctuation before sample yielding, leading to highly dispersed plastic shearing in the entire

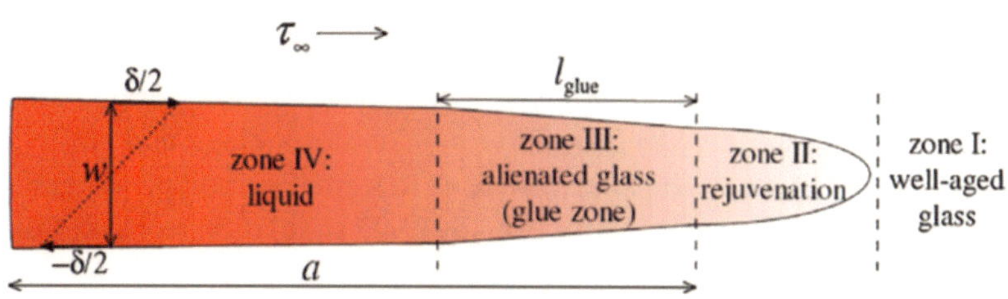

Figure 9. The aged-rejuvenation-glue-liquid (ARGL) shear band model [29].

sample. The transition in deformation mode from highly localized shear banding to nonlocalized plastic deformation was associated with the competition between the yield strength of the material and the critical stress required for the formation of mature SBs in the load-bearing materials. Zhong et al. extracted the SB part of a shear banding deformed MG, and applied tensile stress on the SB sample [33]. The SB tended to have softened tensile behavior and homogenous deformation, with lower tensile stress compared with MG matrix and no obvious localization of atomic strain.

Basing on the above investigation, the cavitation and propagation of SBs seems have a relation with various factors, such as size scale, cooling rate, temperature, composition, etc. Li et al. took MD simulations to investigate the tensile deformation behavior of MG samples with multimillion atoms [34]. They found small rob samples showed remarkable resistance to formation of SBs, and behaved unusual necking phenomenon. Three factors were concluded as contribution to the deformation mode size effects: surface effect, sample loading geometry, and finite sample size. With a further study on the shear banding behavior of double-notched sample, they indicated a critical dimension size about 10–20 nm was needed for the nucleation of SBs [35]. Gao et al. found the size effects in the deformation of Cu-Zr MG. With the model diameter gradually decreasing, the deformation mode of MG evolved from highly localized SB formation to homogenous deformation, but the stress increased significantly during the tensile process [36]. Zhong et al. also had an interest in the size effects, and they found with decreasing film thickness of MG, a transition from the localized deformation to the nonlocalized deformation indeed occurred [37]. Their further study revealed that the critical thickness for this transition was sensitive to the composition, and it was correlated with the average activation energy of the atomic level plastic deformation events [38].

Cheng et al. demonstrated the effects of cooling rate and composition on localization of deformation [11, 12]: with a lower cooling rate, the MG exhibited higher strength but easier trend to strain localization; the MG with composition of $Cu_{64}Zr_{36}$ with high proportion of FI clusters, was more resistant to the initiation of flows but increased propensity to strain localization, while $Cu_{40}Zr_{60}$ was on the opposite side. Zhong et al. utilized this property and created several composites by controlling the layers thickness and numbers of the two MGs, and investigated the deformation behavior of these composites [39]. They found out that MG samples with high layer numbers present obviously nonlocalized deformation behavior, the criterion for the deformation mode change for MGs was suggested as the competition between the elastic energy densities stored and the energy density needed for forming one mature SB in MGs. A further investigation on the annealing effects revealed that the localizing degree of SBs could also be regulated by annealing, with free volume deduction detected during the structural relaxation process [33]. Nanoindentation simulations on binary MG taken by Shi et al. also presented that the SB morphology under indenter had great dependence on indentation rate [40, 41]. At a lower loading rate, SBs showed wing-like morphology and easily propagated to the surface. In the opposite, wedge-like SBs came to formation and penetrated downward to MG matrix at higher loading rate. In our previous MD simulation work on Cu-Zr MG, we also observed more localized SBs under indentation at lower temperature, and more homogeneous deformation morphology at room temperature, which coincided with the brittleness characteristic of most MGs at low temperature [42].

http://dx.doi.org/10.5772/intechopen.76830

The spreading of SBs can also be triggered or suppressed by cyclic loads, and it is relevant to soften or harden behaviors of MGs. Sha et al. performed MD simulations of tension-compression fatigue on $Cu_{50}Zr_{50}$ MG [43]. They observed the initiation and propagation of a major SB throughout the MG sample under cyclic loading. The cycling loads accelerated the accumulation of STZs, and triggered the spreading of SB once the aggregates of STZ reached the critical size for shear banding. Meanwhile, the fully formation of the SB was accompanied with stress drops, and this indicated the soften behavior of MG under uniform axial loads. While Deng et al. observed hardening behavior of Cu-Zr MG under cyclic indentation loads in MD simulations [44]. They indicated the post plastic deformation induced by the earlier indentations suppressed the SB formation by locally stiffening some portions of the original shear banding path, and then a higher load was needed for a secondary path of shear banding yield. Our works demonstrated this hardening behavior could be more severe under higher indentation loads or temperature [42]. In fact, these factors would induce higher plasticity in the deforming region, which formed more complex shear banding morphology and prevented localized SBs formation. These works performed that softening or hardening phenomenon could happen in MGs, due to SBs spreading or suppression under different types of cyclic loads.

5. Deformation behaviors of nanoglass

With highly inhomogeneous deformation dominated by localized SBs, MGs are easily encountered with catastrophically failure. Some pre-deformation process, like cold rolling, leads to a pseudocomposite structure consisting of a softer phase inside pre-induced SBs and a harder phase in the undeformed regions [45, 46]. These microstructural features improve the macroscopic plasticity by promoting the nucleation of secondary SBs and SB branching, as well as by limiting SB propagation due to the intersection of SBs. An alternative method to prevent a major SB development is regulating the volume-interface ratio and density by introducing particle interfaces into the matrix of MG [47]. Such a MG could be produced via cold compaction of glassy nanoparticles, and is therefore called a nanoglass (NG).

According to the definition of nanoglass, Sopu et al. constructed a 3D periodic Cu-Zr nanoglass with an idealized nanostructure consisting of columnar grains with a hexagonal cross section [48]. Applied with tensile stress during a mechanical test simulation, multiple embryonic SBs were formed along the interfaces and eventually started to propagate through the grain interiors in the NGs. Since the elastic energy was released homogeneously in the whole sample, the local energy release was not sufficient to accelerate any SBs; thus, NGs deformed homogeneously in contrast to the MGs, which exhibited localized deformation in one major SB, shown in **Figure 10**. Comparing the strain localization parameters of the NGs and MGs calculated according to Eq. (1), a more homogeneous deformation in the NGs was supported. Moreover, with an annealing treatment to the NGs, the glass particle interfaces seemed have a partial recovery of icosahedral SRO, led to the increase of strain localization.

Adibi et al. constructed a more random distributed NG by involving Poisson-Voronoi tessellation method, using the cast Cu-Zr MG structure as a source of material for the glassy grains

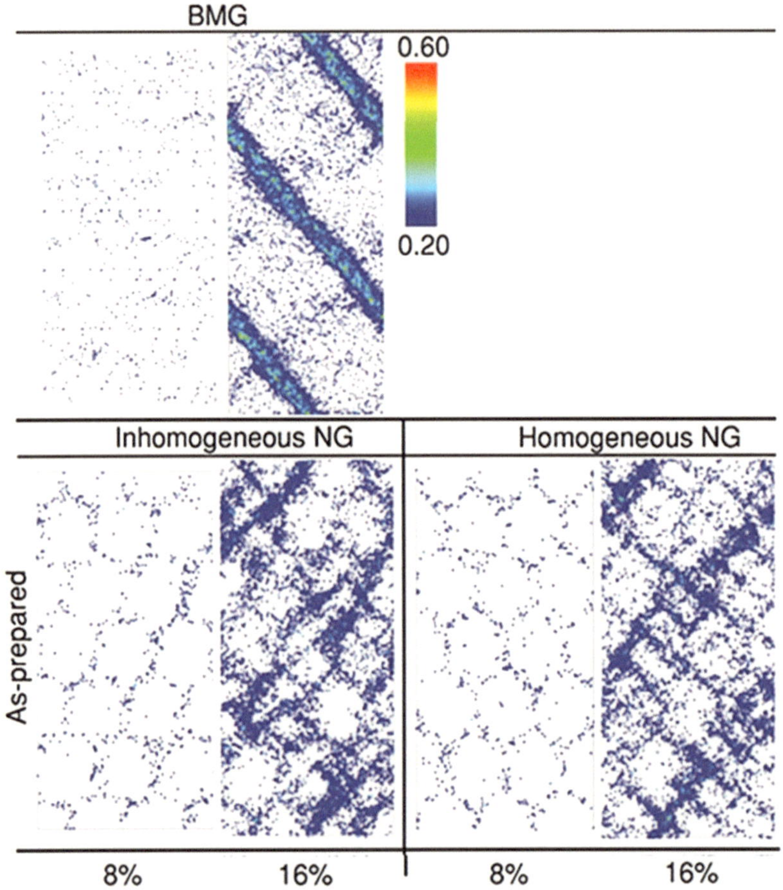

Figure 10. Local atomic shear strain distributions for MG, inhomogeneous NG and homogeneous NG at tensile of 8 and 16% [48].

(shown in **Figure 11**). Three specimens were selected with average glassy grain sizes from 5 to 15 nm to investigate deformation mode transition. The deformation modes of the NGs generally transferred from localized shear banding to homogeneous plastic flow with the grain sizes decreasing, and the fine-grained even became superplastically deformed during tensile testing [49]. The effects of composition and grain size were investigated more specifically by involving Voronoi polyhedral analysis method [50]. They found that the mechanical behavior of NGs was regulated by both the grain boundary thickness and the fraction of atoms at interfaces intrinsically. The mechanical behavior of NGs had a composition dependence similar to their parent MG, while the intrinsic deformation behavior only depended on the grain size, and not affected by the composition.

Sha et al. constructed a sandwich architecture composed of NG and MG [51]. The constructed composite had a higher strength than pure NG, and larger plasticity than MG. The improving of plasticity was contributed by the glass-glass interfaces in NG layers and a compressive residual stress in the MG layer. They indicated a strong ductile MG could be

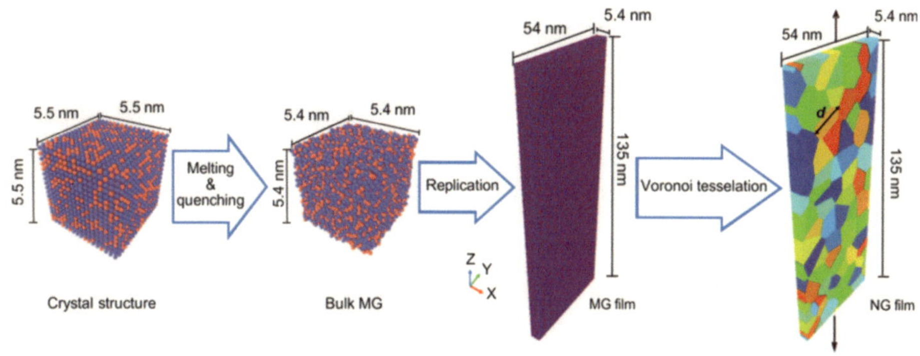

Figure 11. Schematic of the process to generate a NG film sample with Adibi's method [49].

constructed as sandwich architecture, and with a NG surface coated on the surface, MGs could be protected from localized SBs. They further conducted tensile load along the vertical direction to the laminates [52]. The change in the loading direction caused the differences not only in the location of SB initiation but also the critical distance between NG layers for failure mode transition. The MG-NG nanolaminate structure with NG layers closely packed and interfaces oriented parallel to the loading direction was identified as the most effective heterostructure. Adibi et al. constructed another NG-MG nanolaminates by combining layers of MG and NG [53], and further investigated mechanical properties of these nanopillar-shaped nanolaminates with tensile loading simulations. Compared with pure MG and NG, the MG-MG nanolaminate exhibited delayed SB formation and diffused shear banding failure, the NG-MG nanolaminate showed exceptional plasticity to a strain of 0.15 prior to a necking-type failure. These works suggest the MG composites constructed by NG and MG laminates can offer promise for creating structures that combine outstanding strength and ductility.

6. Conclusion and outlook

Throughout the review, we enumerated pronounced work on the structure analysis, the connection of structure with the mechanical properties, shear bands initiation and development. With the application of structural analysis methods, the existence of SRO and MRO in MGs is confirmed. The SRO types and proportions vary with MG compositions and preparation process. Several SRO motifs with high five-fold symmetry structure, as well as their interconnected networks, are found to be great contributing factors to the solid-like characteristics of MG, which means they are closely correlated with mechanical properties and deformation patterns of MGs. The intrinsic dynamics of shear banding events were investigated from the mechanics simulations, which provided convincingly supports for the shear banding theories. Furthermore, we also demonstrated recent MD simulations works on mechanical behaviors and deformation properties of several designed nanoglass. They behaved ductile deformation ability and might be a promising derived material from casted MGs.

These works proved the atomic-level computational simulation plays a significant role in researching work of MG. However, limited by accurate potential descriptions, computational capacity and simulation methods, computational simulations of MG are still unable to build a consistent theoretical frame with experimental research both in time and space. Thus, developing effective multi-scale simulation methods and advanced experimental characterization methods is the most urgent and challenging work for the research of metallic glasses.

Acknowledgements

This work was funded by the National Natural Science Funds for Excellent Young Scholar (51422503), Special Projects for Development of National Major Scientific Instruments and Equipment (2012YQ030075), Fund Guiding on Strategic Adjustment of Jilin Provincial Economic Structure Project (2014Z045), Major Project of Jilin Province Science and Technology development plan (20150203014GX), the special fund project of Jilin provincial industrial innovation (2016C030), Jilin Provincial Middle and Young Scientific and Technological Innovation Talent and Team Project (20170519001JH), Project (2017141) and Project (2017017) Supported by Graduate Innovation Fund of Jilin University.

Conflict of interest

The authors declared that they have no conflicts of interest to this work.

Author details

Hongwei Zhao*, Dan Zhao, Bo Zhu and Shunbo Wang

*Address all correspondence to: hwzhao@jlu.edu.cn

School of Mechanical Science and Engineering, Jilin University, Changchun, Jilin, China

References

[1] Gao YF. An implicit finite element method for simulating inhomogeneous deformation and shear bands of amorphous alloys based on the free-volume model. Modelling & Simulation in Materials Science & Engineering. 2006;**14**:1329-1345. DOI: 10.1088/0965-0393/14/8/004

[2] Yang Q, Mota A, Ortiz M. A finite-deformation constitutive model of bulk metallic glass plasticity. Computational Mechanics. 2006;**37**:194-204. DOI: 10.1007/s00466-005-0690-5

[3] Homer ER, Schuh CA. Mesoscale modeling of amorphous metals by shear transfo-rmation zone dynamics. Acta Materialia. 2009;**57**:2823-2833. DOI: 10.1016/j.actamat.2009.02.035

[4] Li L, Homer ER, Schuh CA. Shear transformation zone dynamics model for metallic glasses incorporating free volume as a state variable. Acta Materialia. 2013;**61**:3347-3359. DOI: 10.1016/j.actamat.2013.02.024

[5] Bernal JD. Geometry of the structure of monatomic liquids. Nature. 1960;**185**:68-70. DOI: 10.1038/185068a0

[6] Sheng HW, Luo WK, Alamgir FM, et al. Atomic packing and short-to-medium-range order in metallic glasses. Nature. 2006;**439**:419. DOI: 10.1038/nature04421

[7] Clarke AS, Jónsson H. Structural changes accompanying densification of random hard-sphere packings. Physical Review E Statistical Physics Plasmas Fluids & Related Interdisciplinary Topics. 1993;**47**:3975. DOI: 10.1103/PhysRevE.47.3975

[8] Honeycutt JD, Andersen HC. Molecular dynamics study of melting and freezing of small Lennard-Jones clusters. Journal of Physical Chemistry. 1987;**91**:4950-4963. DOI: 10.1021/j100303a014

[9] Duan G, Xu D, Zhang Q, et al. Molecular dynamics study of the binary $Cu_{46}Zr_{54}$ metallic glass motivated by experiments: Glass formation and atomic-level structure. Physical Review B. 2005;**71**:224208. DOI: 10.1103/PhysRevB.71.224208

[10] Cheng YQ, Sheng HW, Ma E. Relationship between structure, dynamics, and mechanical properties in metallic glass-forming alloys. Physical Review B. 2008;**78**:014207. DOI: 10.1103/PhysRevB.78.014207

[11] Cheng YQ, Cao AJ, Ma E. Correlation between the elastic modulus and the intrinsic plastic behavior of metallic glasses: The roles of atomic configuration and alloy composition. Acta Materialia. 2009;**57**:3253-3267. DOI: 10.1016/j.actamat.2009.03.027

[12] Ward L, Miracle D, Windl W, et al. Structural evolution and kinetics in Cu-Zr metallic liquids from molecular dynamics simulations. Physical Review B. 2013;**88**:134205. DOI: 10.1103/PhysRevB.88.134205

[13] Cheng YQ, Ma E, Sheng HW. Atomic level structure in multicomponent bulk metallic glass. Physical Review Letters. 2009;**102**:245501. DOI: 10.1103/PhysRevLett.102.245501

[14] Liu XJ, Xu Y, Hui X, et al. Metallic liquids and glasses: Atomic order and global packing. Physical Review Letters. 2010;**105**:155501. DOI: 10.1103/PhysRevLett.105.155501

[15] Liang YC, Liu RS, Mo YF, et al. Influence of icosahedral order on the second peak splitting of pair distribution function for $Mg_{70}Zn_{30}$ metallic glass. Journal of Alloys & Compounds. 2014;**597**:269-274. DOI: 10.1016/j.jallcom.2014.01.052

[16] Nelson DR, Spaepen F. Polytetrahedral order in condensed matter. Solid State Physics. 1989;**42**:1-90. DOI: 10.1016/S0081-1947(08)60079-X

[17] Tang C, Wong CH. A molecular dynamics simulation study of solid-like and liquid-like networks in $Zr_{46}Cu_{46}Al_8$ metallic glass. Journal of Non-Crystalline Solids. 2015;**422**:39-45. DOI: 10.1016/j.jnoncrysol.2015.05.003

[18] Liu YH, Wang D, Nakajima K, et al. Characterization of nanoscale mechanical heterogeneity in a metallic glass by dynamic force microscopy. Physical Review Letters. 2011; **106**:125504. DOI: 10.1103/PhysRevLett.106.125504

[19] Cheng YQ, Ma E. Atomic-level structure and structure–property relationship in metallic glasses. Progress in Materials Science. 2011;**56**:379-473. DOI: 10.1016/j.pmatsci.2010.12.002

[20] Wakeda M, Shibutani Y, Ogata S, et al. Relationship between local geometrical factors and mechanical properties for cu–Zr amorphous alloys. Intermetallics. 2007;**15**:139-144. DOI: 10.1016/j.intermet.2006.04.002

[21] Cao AJ, Cheng YQ, Ma E. Structural processes that initiate shear localization in metallic glass. Acta Materialia. 2009;**57**:5146-5155. DOI: 10.1016/j.actamat.2009.07.016

[22] Feng S, Qi L, Wang L, et al. Atomic structure of shear bands in $Cu_{64}Zr_{36}$ metallic glasses studied by molecular dynamics simulations. Acta Materialia. 2015;**95**:236-243. DOI: 10.1016/j.actamat.2015.05.047

[23] Ju SP, Huang HH, Wu TY. Investigation of the local structural rearrangement of $Mg_{67}Zn_{28}Ca_5$ bulk metallic glasses during tensile deformation: A molecular dynamics study. Computational Materials Science. 2015;**96**:56-62. DOI: 10.1016/j.commatsci.2014.09.005

[24] Lee M, Lee CM, Lee KR, et al. Networked interpenetrating connections of icosahedra: Effects on shear transformations in metallic glass. Acta Materialia. 2011;**59**:159-170. DOI: 10.1016/j.actamat.2010.09.020

[25] Jeong HY, Li XW, Yee AF, et al. Slip lines in front of a round notch tip in a pressure-sensitive material. Mechanics of Materials. 1994;**19**:29-38. DOI: 10.1016/0167-6636(94)90035-3

[26] Su C, Anand L. Plane strain indentation of a Zr-based metallic glass: Experiments and numerical simulation. Acta Materialia. 2006;**54**:179-189. DOI: 10.1016/j.actamat.2005.08.040

[27] Argon AS. Plastic deformation in metallic glasses. Acta Metallurgica. 1979;**27**:47-58. DOI: 10.1016/0001-6160(79)90055-5

[28] Ogata S, Shimizu F, Li J, et al. Atomistic simulation of shear localization in Cu–Zr bulk metallic glass. Intermetallics. 2006;**14**:1033-1037. DOI: 10.1016/j.intermet.2006.01.022

[29] Shimizu F, Ogata S, Li J. Yield point of metallic glass. Acta Materialia. 2006;**54**:4293-4298. DOI: 10.1016/j.actamat.2006.05.024

[30] Shimizu F, Ogata S, Li J. Theory of shear banding in metallic glasses and molecular dynamics calculations. Materials Transactions. 2007;**48**:2923-2927. DOI: 10.2320/matertrans. MJ200769

[31] Lewandowski JJ, Greer AL. Temperature rise at shear bands in metallic glasses. Nature Materials. 2006;**5**:15-18. DOI: 10.1038/nmat1536

[32] Zhou HF, Zhong C, Cao QP, et al. Non-localized deformation in metallic alloys with amorphous structure. Acta Materialia. 2014;**68**:32-41. DOI: 10.1016/j.actamat.2014.01.003

[33] Zhong C, Zhang H, Cao QP, et al. Deformation behavior of metallic glasses with shear band like atomic structure: A molecular dynamics study. Scientific Reports. 2016;**6**:30935. DOI: 10.1038/srep30935

[34] Li QK, Li M. Free volume evolution in metallic glasses subjected to mechanical deformation. Materials Transactions. 2007;**48**:1816-1821. DOI: 10.2320/matertrans.MJ200785

[35] Li QK, Li M. Assessing the critical sizes for shear band formation in metallic glasses from molecular dynamics simulation. Applied Physics Letters. 2007;**91**:231905. DOI: 10.1063/1.2821832

[36] Gao LK, Zhao FL, Xu N, et al. Size effects: The relation to the percentage of atoms that participate in the deformation of ZrCu metallic glass. Journal of Spectroscopy. 2014;**2014**:1-5. DOI: 10.1155/2014/627679

[37] Zhong C, Zhang H, Cao QP, et al. The size-dependent non-localized deformation in a metallic alloy. Scripta Materialia. 2015;**101**:48-51. DOI: 10.1016/j.scriptamat.2015.01.015

[38] Zhong C, Zhang H, Cao QP, et al. On the critical thickness for non-localized to localized plastic flow transition in metallic glasses: A molecular dynamics study. Scripta Materialia. 2016;**114**:93-97. DOI: 10.1016/j.scriptamat.2015.12.012

[39] Zhong C, Zhang H, Cao QP, et al. Non-localized deformation in Cu-Zr multi-layer amorphous films under tension. Journal of Alloys & Compounds. 2016;**678**:410-420. DOI: 10.1016/j.jallcom.2016.03.305

[40] Shi Y, Falk ML. Simulations of nanoindentation in a thin amorphous metal film. Thin Solid Films. 2007;**515**:3179-3182. DOI: 10.1016/j.tsf.2006.01.032

[41] Shi Y, Falk ML. Stress-induced structural transformation and shear banding during simulated nanoindentation of a metallic glass. Acta Materialia. 2007;**55**:4317-4324. DOI: 10.1016/j.actamat.2007.03.029

[42] Zhao D, Zhao HW, Zhu B, et al. Investigation on hardening behavior of metallic glass under cyclic indentation loading via molecular dynamics simulation. Applied Surface Science. 2017;**416**:14-23. DOI: 10.1016/j.apsusc.2017.04.125

[43] Sha ZD, Qu SX, Liu ZS, et al. Cyclic deformation in metallic glasses. Nano Letters. 2015;**15**:7010-7015. DOI: 10.1021/acs.nanolett.5b03045

[44] Deng C, Schuh CA. Atomistic mechanisms of cyclic hardening in metallic glass. Applied Physics Letters. 2012;**100**:4067. DOI: 10.1063/1.4729941

[45] Lee MH, Lee KS, Das J, et al. Improved plasticity of bulk metallic glasses upon cold rolling. Scripta Materialia. 2010;**62**:678-681. DOI: 10.1016/j.scriptamat.2010.01.024

[46] Cao QP, Liu JW, Yang KJ, et al. Effect of pre-existing shear bands on the tensile mechanical properties of a bulk metallic glass. Acta Materialia. 2010;**58**:1276-1292. DOI: 10.1016/j.actamat.2009.10.032

[47] Jing J, Krämer A, Birringer R, et al. Modified atomic structure in a Pd-Fe-Si nanoglass: A Mössbauer study. Journal of Non-Crystalline Solids. 1989;**113**:167-170. DOI: 10.1016/0022-3093(89)90007-0

[48] Şopu D, Ritter Y, Gleiter H, et al. Deformation behavior of bulk and nanostructured metallic glasses studied via molecular dynamics simulations. Physical Review B. 2011;**83**. DOI: 10.1103/PhysRevB.83.100202

[49] Adibi S, Sha ZD, Branicio PS, et al. A transition from localized shear banding to homogeneous superplastic flow in nanoglass. Applied Physics Letters. 2013;**103**:. 211905-211905-5. DOI: 10.1063/1.4833018

[50] Adibi S, Branicio PS, Zhang YW, et al. Composition and grain size effects on the structural and mechanical properties of CuZr nanoglasses. Journal of Applied Physics. 2014;**116**:2744. DOI: 10.1063/1.4891450

[51] Sha ZD, He LC, Pei QX, et al. The mechanical properties of a nanoglass/metallic glass/nanoglass sandwich structure. Scripta Materialia. 2014;**83**:37-40. DOI: 10.1016/j.scriptamat. 2014.04.009

[52] Sha ZD, Branicio PS, Lee HP, et al. Strong and ductile nanolaminate composites combining metallic glasses and nanoglasses. International Journal of Plasticity. 2017;**90**:231-241. DOI: 10.1016/j.ijplas.2017.01.010

[53] Adibi S, Branicio PS, Ballarini R. Compromising high strength and ductility in nanoglass–metallic glass nanolaminates. RSC Advances. 2016;**6**:13548-13553. DOI: 10.1039/C5RA24715B

Section 4

Metallic Glass Matrix Composites

Chapter 4

Metallic Glass Matrix Composites

Wei Guo

Additional information is available at the end of the chapter

http://dx.doi.org/10.5772/intechopen.76526

Abstract

In this chapter, we will firstly introduce the metallic glass and its deformation characteristics briefly. Then we will focus on the fabrication and mechanical properties of metallic glass matrix composites, including ex-situ particle/fiber reinforced, in-situ precipitated phase reinforced and etc. The alloy systems will cover from Zr-based, Ti-based to Mg-based and etc. We will also introduce the latest research on both new reinforcing phases and fabrication processes, including porous particles, shape memory phases, novel dealloying in metallic melt method, selective phase leaching method and so on. The microstructures, mechanical properties of each kind of composites, as well as the optimization methods, will be discussed in detail.

Keywords: metallic glass matrix composite, mechanical property, plasticity, fabrication process, shear band

1. Introduction

1.1. What is metallic glass?

The metallic glass or bulk metallic glass (BMG) is the alloy without crystals, or so-called non-crystalline alloys. In BMGs, the microstructure is so-called amorphous state, referring to the long-range disordered structures between atoms inside a material. The amorphous materials can be produced by non-crystalized cooling from melting state or vapor deposition, mechanical alloying methods, etc. To date, the amorphous materials occupy a large proportion in nature materials, from conventional oxide glass to amorphous semi-conductor, then to amorphous metals or bulk metallic glasses. The amorphous materials have been very important engineering materials to support the modern economy, as well as economic and social developments. Besides the daily-used glassy materials, in high-tech fields, amorphous materials have been also applied a lot to optical communication, laser technology, new solar battery, power transmission materials and so on.

Unlike the conventional oxide glasses, the amorphous alloys or metallic glasses possess metallic bond between atoms instead of covalent bond. Thus, the characteristics related with metals are maintained, such as opacity, good toughness, etc. We can say the amorphous structure is faultless for the lack of dislocations or grain boundaries. We can also say the amorphous structure or random-arranged structure is full of defects because you can find no periodicity in it.

Even though both amorphous alloys and bulk metallic glasses are noncrystalline materials, which are obtained from rapid cooling from liquid state, hindering the crystallization kinetics [1]. The high rate of heat transfer required to prevent crystallization often limits these noncrystalline materials to thin samples or ribbon-shaped samples. These noncrystalline materials are called amorphous alloys. Recently, bulk metallic glasses with slower nucleation kinetics in undercooled liquids have been processed by conventional casting at cooling rate of 10^{-1}–10^{-2} K/s [2–6]. The critical size can be larger than 1 mm rods; these "bulk" noncrystalline alloys are called bulk metallic glasses. **Figure 1** illustrates the conditions for processing both the more recently developed bulk amorphous alloys as well as traditional metallic glass alloys developed before 1990 [4]. The plot correlating critical cooling rate and maximum sample thickness as a function of the reduced glass transition temperature (T_g/T_m) shows a clear tendency for the glass-forming ability to increase with increasing T_g/T_m, as such these alloys have lower critical cooling rates and larger possible bulk cross-sectional dimensions.

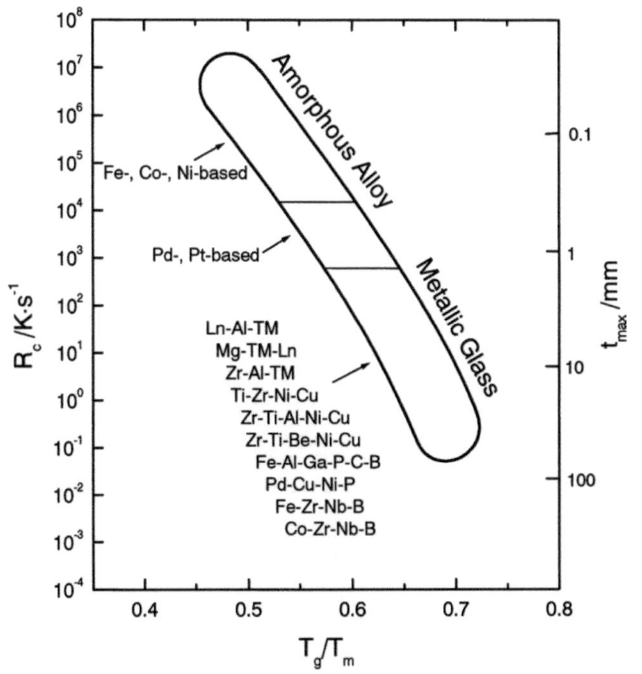

Figure 1. Plot correlating critical cooling rate (R_c), maximum sample thickness (t_{max}), and reduced glass transition temperature (T_g/T_m) for bulk metallic alloys illustrating conditions for processing both the more recently developed bulk amorphous alloys as well as ordinary amorphous alloys developed before 1990 [4].

The criteria [7–10] for slow crystallization kinetics, a stabilized supercooled liquid and high glass-forming ability, resulting in the formation of bulk metallic glasses, have been shown to include:

1. Multi-component alloys of increased complexity and size of crystal unit cells such that the energetic advantage of an ordered structure is reduced by increasing the configurational entropy of the supercooled liquid phases.
2. Atomic radius mismatch between elements, which leads to higher packing density and smaller free volume, requiring a greater volume increase for crystallization, as well as limiting the solubility of these atoms in crystalline states.
3. Negative heat of mixing between the elements, which increases the energy barrier at the solid–liquid interface and accelerates atomic diffusivity, thus slowing local atomic rearrangements and crystal nucleation rate, thereby extending the supercooled liquid temperature.
4. Alloy composition close to deep eutectic, which forms a liquid stable at low temperatures that can freeze into the glassy state.

1.2. Why study metallic glass?

The research and developments of BMGs indicate that compared with traditional crystalline materials, BMGs have an advantage in usability. The main points are as follows:

1. Better mechanical properties such as high yielding strength, large elastic strain limit, mainly perfect elasticity before yielding, mainly perfect plasticity after yielding, no work hardening, high fatigue resistance and high abrasive resistance. With the developments of BMGs, the ultimate strength of metallic materials is renewed again and again. The strength of Mg-based BMGs has increased from 600 to 800 MPa [11]. The strength of Cu-based BMGs is over 2000 MPa [12]. Especially for Co-Fe-Ta-B alloy, the strength is over 5000 MPa [12], which sets up a record in natural world.
2. Good processability. Near glass transition temperature (T_g), the elongation of La-Al-Ni BMGs can be 15,000% [13]. Other BMGs also show super-plasticity to varying degrees, thus according to different application, BMGs can be manufactured into micro- or nano-level by machining deformation.
3. Better corrosion resistance against many kinds of medium. The corrosion resistance of Fe-Cr-Mo-B-P BMGs is 10,000 times higher than conventional stainless steel and can be used in much severed environments [14].
4. Good physical properties such as soft and hard magnetism, unique expansive quantity. For example, the saturation magnetization of Fe-based amorphous alloys can be over 1.5 T and coercivity is lower than 1 A/m^2 [15]. When some BMGs are annealed to form nanocrystalline alloys, better soft or hard magnetism can be obtained, which are considered as excellent substitute for conventional materials.

For the better physical, chemical, mechanical properties and precision shaping abilities of BMGs than conventional materials, BMGs have shown important application value in aerospace device, precision machine, information technology and so on. The researches of BMGs have attracted a lot of attentions from physical, chemical and material scientists.

1.3. Room-temperature brittleness of metallic glass

Because of the metallic bonding in amorphous alloys, strain can be accommodated at the atomic level through changes in neighborhood; atomic bonds can be broken and reformed at the atomic scale. However, unlike crystalline metals and alloys, metallic glasses do not exhibit long-range translational symmetry. Thus, the deformation mechanisms such as dislocations, which allow changes in atomic neighborhood at low energies or stresses, do not exist in metallic glasses. The local rearrangement of atoms in metallic glasses is a relatively high-energy or high-stress process. The exact nature of local atomic motion in deforming metallic glasses is not fully understood, although there is general consensus that the fundamental unit must be a local rearrangement of atoms accommodating the shear strain. An example of such a local rearrangement is depicted in the two-dimensional schematic of **Figure 2a**, originally proposed by Argon and Kuo [16] on the basis of an atomic-analog bubble-raft model, called a "shear transformation zone" (STZ) [17–20]. The STZ is essentially a local cluster of atoms that undergoes an inelastic shear distortion from one relatively low energy configuration to a second such configuration, crossing an activated configuration of higher energy and volume. The STZs are common to deformation of all amorphous metals, although details of the structure, size and energy scales of STZs may vary from one glass to the other. In a metallic glass body experiencing uniform stress, the STZ that is activated first is selected from among many potential sites on the basis of energetics, which vary with the local atomic arrangements [21–23]. The continued propagation of the applied shear strain occurs when one STZ creates a localized distortion of the surrounding material, which triggers the formation of large planar bands of STZs along the maximum shear stress plane, or so-called "shear bands", as shown in **Figure 2b**. For most BMGs, the deformation occurs in homogeneous through plastic strains

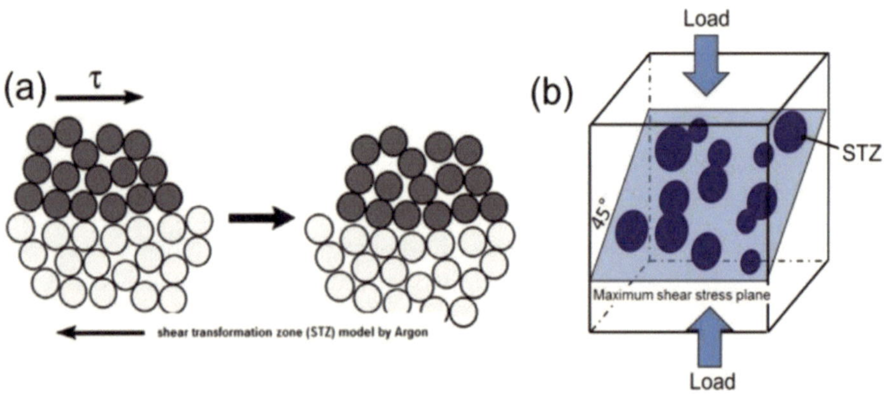

Figure 2. Schematic illustrating of (a) "shear transformation zone" in which strain accommodation occurs through localized cluster of atoms undergoing intense distortion and (b) shear band formation along maximum shear stress plane.

concentrated in localized shear bands at room temperature. Once a shear band initiates, the propagation of it can be very fast (~1000 m/s), thus, the BMGs fracture catastrophically after elastic deformation. Therefore, the room-temperature brittleness, especially under the uniaxial compression or tension, has been one fatal problem for the wide application of BMGs.

1.4. How to overcome the problem?

To date, there are several ways that have been developed to improve the room-temperature plasticity of BMGs, including intruding a secondary phase to develop a composite microstructure, surface coating, composition adjustment to induce intrinsic heterogeneity, severe plastic deformation such as shot peening and high pressure torsion [24–27]. Although introducing a secondary phase to make a composite structure seems to be most primal methods, it is reported that composite structure is one of the most efficient method and very easy to realize. Furthermore, there are many combinations of amorphous matrix and reinforcing phases, giving an infinite possibility to improve the mechanical properties of BMGs.

For now, there are mainly two ways to introduce the secondary phases, ex-situ direct adding and in-situ precipitation. For ex-situ method, the various combinations of secondary particle or fiber and the amorphous matrix makes the fabrication process easier to design. But the interface bonding between the secondary phase and the matrix is not strong because of the formation of surface oxide layers, which degrades the mechanical properties of them. For in-situ method, even though the interface bonding is stronger than those ex-situ composites because the secondary phases are intrinsically formed in the melt during cooling, but the fabrication process is very difficult to design. Furthermore, for ex-situ method, the size and volume fraction of secondary phases can be easily controlled by using various sized particles or fibers with various amounts when adding. However, for in-situ method, it is difficult to optimize the microstructures because the optimization process is related with the composition adjustment. Hereafter, we will introduce the researches on both ex-situ and in-situ BMGMCs, including their composite structure and mechanical properties.

2. Ex-situ BMGMCs

The selection of ex-situ secondary phases includes fibers, particles, pores and porous particles. Hereafter, we will introduce the microstructures and mechanical properties of each kind of ex-situ BMGMCs.

2.1. Fiber-reinforced BMGMCs

The fiber-reinforced BMGMCs mainly focus on tungsten fiber, steel fiber and carbon fiber [28–34]. Dandliker et al. have firstly fabricated the tungsten and carbon-steel continuous wire reinforced Zr-based BMGMCs by quenching the metallic melt to a glass after infiltrating the reinforcement [28]. The continuous long fibers in the glass matrix can efficiently hinder the propagation of main shear bands, improving the plastic strain from 0% of monolithic BMG to over 2% of those reinforced with steel wires. Kim et al. have successfully fabricated carbon

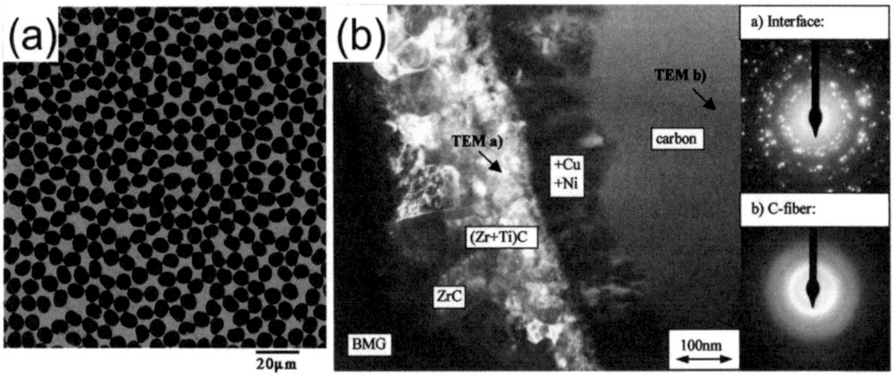

Figure 3. (a) Backscattered SEM image of carbon fiber reinforced bulk metallic glass composite; (b) dark field TEM of the interfacial region between a carbon fiber and the matrix.

fiber reinforced Zr-based BMGMCs by infiltrating the alloy melt to the bundle of carbon fibers in a quartz tube which are cleaned and preheated [30]. **Figure 3a** shows the backscattered SEM image of the carbon fiber reinforced composite. The carbon fibers are uniformly distributed in the matrix and the matrix appears to be uniform and free of heterogeneity. The volume fraction of carbon fibers is about 50% and the diameter is about 5 μm. They also found that a carbide reaction zone is formed surrounding the carbon fibers, as shown in **Figure 3b**, starting from the carbon fibers, a diffusion zone of Ni, Cu within the fiber, a crystallize reaction zone of (Zr + Ti)C and ZrC, to the BMG matrix. Qiu et al. cast the sample in a resistive furnace by melting the ingots in an evacuated quartz tube packed with the tungsten fibers, followed by pressure infiltration [34]. After pressurization the tube was quenched in a supersaturated brine solution. The nominal diameter of the fibers is 250 μm. The volume fraction of the fibers varies from 10 to 70%.

During compression test, unlike the catastrophic fracture of monolithic BMG, the composites reinforced with tungsten fibers shows yielding and plastic deformation. The yielding strength also increases with a higher volume fraction of fibers. They also found that the failure mode changes with various volume fractions of fibers. The monolithic glass fractures on 45° planes. As the volume fraction of fiber increases, failure mode shifts from shear to localized buckling and tilting.

2.2. Particle-reinforced BMGMCs

The particle-reinforce BMGMCs includes ceramic particles, metal particles and the matrix includes Zr-based, Ti-based and Mg-based alloy system [35–43].

Choi-Yim and Johnson have firstly introduced ceramic particle such as SiC, WC and TiC, and the metal particles W and Ta into Zr-based and Cu-based BMG matrix [35]. A mixture of the pre-alloyed metallic glass forming elements and secondary phase material are combined by induction melting the glass forming alloy together with the solid secondary phase material on a water-cooled copper boat under a Ti-gettered argon atmosphere. The volume fractions of particles range from 5 to 30% and the sizes of the particles vary between 20 and 80 μm. **Figure 4**

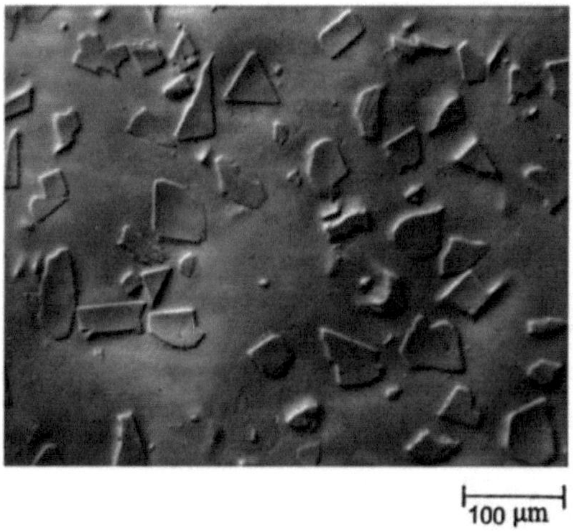

Figure 4. An optical micrograph showing uniformly distributed WC particles in the matrix.

shows the uniformly distributed WC particles in the metallic glass matrix with the volume fraction of 10%. The matrix composition is chosen for several reasons. A relatively low melting temperature suppresses the chemical interactions between the reinforcement particles and the glass. A low glass transition temperature decreases differential thermal stresses which arise between the reinforcement and the matrix during freezing and cooling.

Zhang et al. have introduced Ta particles into Zr-Cu-Al-Ag BMG matrix, the average size of Ta particles is about 40 μm and the volume fraction varies from 5 to 20% [36]. The composite is prepared by induction melting the Zr-Cu-Al-Ag alloy together with Ta powder in a quartz tube and subsequently injecting through a nozzle into a copper mold. **Figure 5a** shows the SEM images of the as-cast BMGMCs containing 10% Ta, consisting of homogeneously dispersed particles embedded in the metallic glass matrix. **Figure 5b** shows the compressive stress-strain curves of the composites containing 5–20% Ta. The monolithic glassy alloy fails immediately after yielding at 1885 MPa. However, the composites exhibit apparent work hardening and plastic strain. For 10% Ta-containing composite, the yield strength, fracture strength and plastic strain are 1717, 2600 MPa and 31%, respectively. The composite containing 20% Ta shows no plasticity which may originate from the crystallization of the matrix. Ta particles play an important role in the initiation and propagation of the multiple shear bands. The differences in Young's modulus between Ta particles and glassy matrix generates high-stress concentration occurs at the interfaces, which promotes the initiation of shear bands at the interface prior to the formation of shear bands on the maximum shear surface.

Pan et al. have added Nb particles into Mg-based BMG matrix to fabricate the ex-situ particle-reinforced BMGMCs [43]. The size of Nb particles is 20–50 μm, the volume fraction varies from 4 to 8%. The particles are added during inductively melting the master alloy. The composite alloy is remelted by induction in a quartz tube and injected with argon pressure into copper molds. **Figure 5c** shows the SEM micrograph of the cross-sectional surface of the composite with 8% Nb particles, which exhibits the uniform distribution of the particles

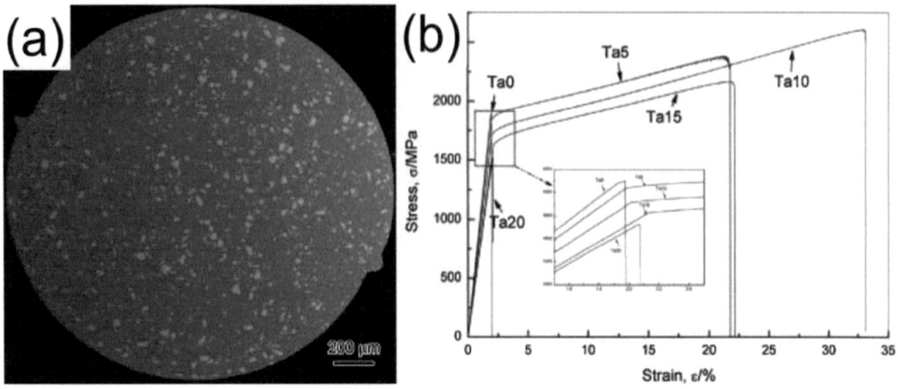

Figure 5. (a) SEM image of the as-cast BMGMC containing 10% Ta; (b) compressive stress-strain curves of the BMGMCs containing 0–20% Ta; (c) SEM image of the composite containing 8% Nb (inset shows the XRD pattern); (d) compressive stress-strain curves for single phase BMG and Nb-containing composite (4 and 8%).

without interfacial reaction. **Figure 5b** shows stress-strain curves for monolithic BMG and Nb-containing composite with 4 and 8% Nb particles. The BMG fails just after the elastic limit of 2%, but the composites yield at about 900 MPa and exhibit significantly plastic strain as well as work hardening and softening. The overall engineering plastic strain is determined to be about 12.1% for 8% Nb-containing composite. Ductile Nb particles serve as obstacles to impede shear bands propagation. When encountering an Nb particle, the shear band has to be either blocked or bypass around the particle due to the strong bonding between the particles and the matrix. Furthermore, if the sear deformation travels into the ductile Nb particles, the particle can dramatically plastic deform by dislocation mode to absorb the shear strain and prevent the catastrophic failure from taking place by the free propagation of unstable shear bands.

2.3. Other kind of ex-situ BMGMCs

Besides fiber and particle reinforcement, recently, some new kind of secondary phases have been introduced to BMG matrix, such as pores and porous particles [43–46].

Wada et al. introduced pores into Pd-based BMG matrix [44]. The master alloy melts are subjected to four distinct hydrogenation treatments at 853 K in tubes of fused silica: (i) 12 h at 1 MPa, then oil quench, (ii) the same, then water quench; (iii) 3 h at 4 MPa, then oil quench, and (iv) the same, then water quench to form porous rods. The porosities of BMG rods are calculated to be 1.7–3.7% from their density. The pore size is observed to be 20–30 μm, as shown in **Figure 6a**. The stability of the fine uniform pore distribution during casting follows from the high viscosity of the melt. Compressive stress-strain curves are strongly affected by porosity, as shown in **Figure 6b**. The porous alloy with the highest porosity of 3.7% shows the plastic strain over 18%, greatly enhanced compared to the pore-free BMG. But during tension, no plasticity can be observed. The shear band pattern is affected by the pores acting as stress concentrators. The pores are comparable in radius with the notch roots giving enhanced toughness, and are expected to induce extensive shear banding.

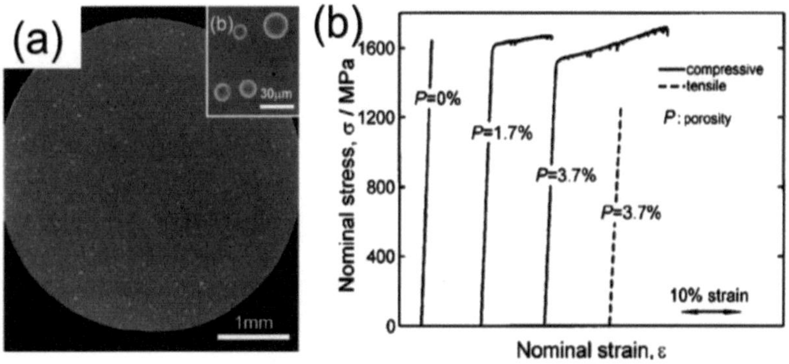

Figure 6. (a) Pore distributed BMG rods, water quenched from a melt held at 853 K for 3 h under 4 MPa hydrogen; (b) tensile and compressive stress-strain curves of pore reinforced BMGMCs with different porosities.

Jang et al. have introduced porous Mo particles into Mg-Cu-Gd-Ag BMG matrix [45]. During master alloy melting, high purity porous Mo particles with a spherical shape and a size of 20–70 μm are added in the matrix alloy. The volume fraction of porous Mo particles ranges from 10 to 25%. For the introduced porous particles, the overall microstructure of the composite is separated into larger-scale compartment, ~50 μm between Mo particles, and the fine-scale compartment, 1–5 μm within one porous Mo particle, as shown in **Figure 7a** and **b**. The composites containing 20–25 vol% porous Mo particles exhibit superior mechanical performance with ultimate compression stress up to 0.95 and 1.1 GPa and plastic strain up to 10%, as shown in **Figure 7c**. Unlike solid particles, the crack propagation in the present composite is arrested by the porous Mo particles. For the porous nature of the reinforcing phase, the numbers of particles are calculated to be 1.4 times higher than those with solid ones, i.e., the mean interspacing of the porous Mo particles is less than that of the solid Mo ones. This should favor to the confining the shear-banding behavior, thus enable more halting propagation of the shear bands. The porous Mo particles separates and restricts the highly localized shear banding into many isolated small regions, and can confine lots of microsized compartments of the matrix within porous particles, which results in the formation of multiple shear bands within or around the porous particles, promoting the deformation to distribute more uniformly across the sample.

Guo et al. have used an original method, so-called top-down process, to fabricate porous NiTi shape memory alloy (SMA) powders [46]. In this process, the multiphase precursor powder of Ni-Ti-Gd is firstly produced with B2-NiTi and Ni-Gd phase, then by leaching in nitric acid solution to remove Ni-Gd phase and leaving pore in the powder, as shown in **Figure 8a**. The size and interspacing between NiTi within one porous particle are as small as 200 nm. The porous NiTi powders are subsequently added to Mg-Cu-Gd-Ag glass former liquid to fabricate the BMGMC, as shown in **Figure 8b**, with volume fraction ranged from 5 to 20%. The porous particles are homogeneously distributed in the glassy matrix. The composites containing 20 vol% porous NiTi particles exhibits the best mechanical properties, including a true plastic strain of up to 10.6% and a fracture stress of up to 1173 MPa, as shown in **Figure 8c**. Similar to porous Mo, many microcracks are confined in the inter-particle regions and should result from local shear banding within the amorphous matrix. Furthermore, compared with solid particles, the porous particles can generate

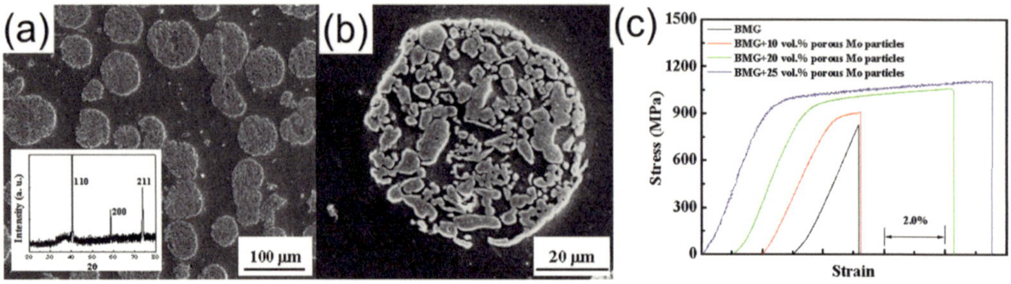

Figure 7. (a) SEM observation of the porous Mo particles in the BMG matrix, with the inserted XRD pattern; (b) an enlarged image of a single porous Mo particle; (c) representative room-temperature compressive engineering stress-strain curves for the BMGMCs.

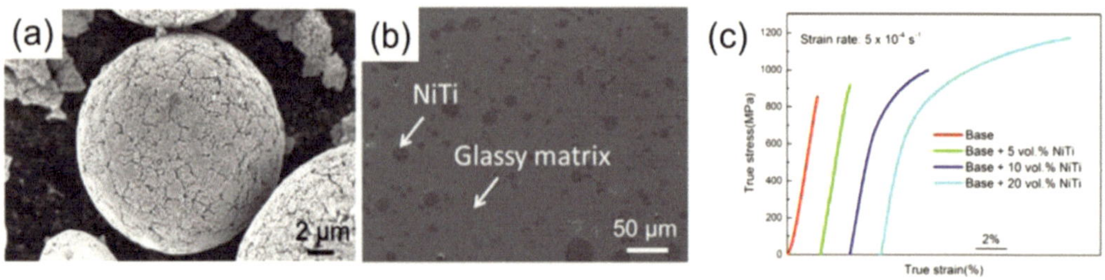

Figure 8. (a) Porous NiTi powder by top-down process; (b) SEM images of BMGMCs containing 20 vol% porous particles; (c) compressive true stress-strain curves for monolithic base BMG and BMGMCs with various volume fraction of porous NiTi addition.

more interfaces, which makes the yield strength follows the load-bearing mode even with low volume fraction of particles. This composite also shows very obvious work hardening behavior, which is considered to originate from the stress-induced martensitic transformation of B2-NiTi phase, which is very attractive and different from those conventional metal or ceramic particles.

3. In-situ BMGMCs

For in-situ BMGMCs, the fabrication process is very important and difficult to design, such as how to induce in-situ precipitated phase, how to control the volume fraction and size of secondary phase. For different alloy systems, the process and reinforcing phases are quite different, thus, this part will be divided by alloy systems, not by reinforcing phases.

3.1. Zr-based BMGMCs

β-Zr phase is one of the in-situ reinforcements in Zr-based BMGMCs that has been studied a lot. Hays et al. have reported the development of β phase reinforced Zr-Ti-Cu-Ni-Be-Nb BMGMCs [47]. Both XRD patterns and SEM images suggest the precipitation of β phase during rapid quenching. The β phase is in the dendritic and distributed in the glassy matrix homogeneously, as shown in **Figure 9a**. The volume fraction of β phase is estimated to be ~25%. The dendritic

http://dx.doi.org/10.5772/intechopen.76526

structures are characterized by primary dendrite axes with lengths of 50–150 μm and radius of about 1.5–2 μm. Regular patterns of secondary dendrite arms with spacing of 6–7 μm can be observed from SEM images. This composite shows about 5% plastic strain under three point bending. **Figure 9b** shows that the shear bands propagate preferentially through many successive dendrite arms, occasionally initiate or terminate within the arms, and clearly propagate as localized bands through the β-phase arms. This composite also shows good plasticity during compression, as shown in **Figure 9c**. It yields at 1.3 GPa when the β phase yields and deforms, and shear band patterns develop, as the glassy matrix is locally loaded beyond its critical shear stress. The plastic strain is over 6%. The composite even shows about 5% plastic strain during tension. Clear necking and deformation can be observed. The dendritic microstructure of the β phase acts to seed the initiation of organized shear band patterns, confines the propagation of individual shear bands to domains having a spatial scale of the order of the primary dendritic axes length, and lead to shear band spacing which is related to the dendrite arm spacing.

Another important in-situ secondary phase in Zr-based BMGMCs is refractory metal phase. Fan et al. have introduced in-situ Ta-rich precipitates in Zr-Cu-Al-Ni-Ta BMGMCs [48]. For the high melting temperature of Ta, there are two steps arc-melting during master alloy preparation. Firstly, Zr-Ta ingot is fabricated which forms solid solutions of Zr-Ta, and remaining elements are subsequently mixed with Zr-Ta ingots. The microstructure of the as-cast composite samples consists of both glassy matrix and Ta-rich particles with an average size of 10–30 μm, as shown in **Figure 10a**. The particles are oblong in shape and do not appear to possess a dendritic structure, they distribute homogeneously among the matrix and the volume faction is about 4%. This composite yields at 1.7 GPa and exhibits apparent work hardening and significant plastic strain, as shown in **Figure 10b**. The elastic incompatibility between the particles and the matrix introduces stress concentrations which may promote shear band initiation. The particles may also impede shear band propagation. Guo et al. have applied dealloying in metallic melt method to further optimize the microstructure and mechanical properties of Ta-rich phase reinforced BMGMCs [49]. The dealloying in metallic melt phenomenon occurs when immerging Zr-Ta solid solution precursor in Cu-Al-Ni melt. For the negative enthalpy of mixing between Zr and Cu-Al-Ni and positive enthalpy of mixing between Ta and Cu-Al-Ni, Zr is gradually selectively leached from precursor ingot to

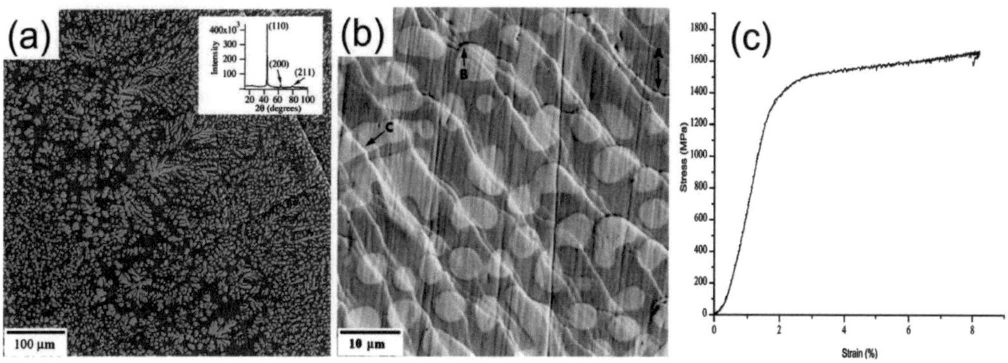

Figure 9. (a) SEM image of in-situ β-Zr reinforced BMGMCs (inset: XRD patterns); (b) shear band patterns array from compressive failure region of bend test sample; (c) compressive stress strain curve for the composite.

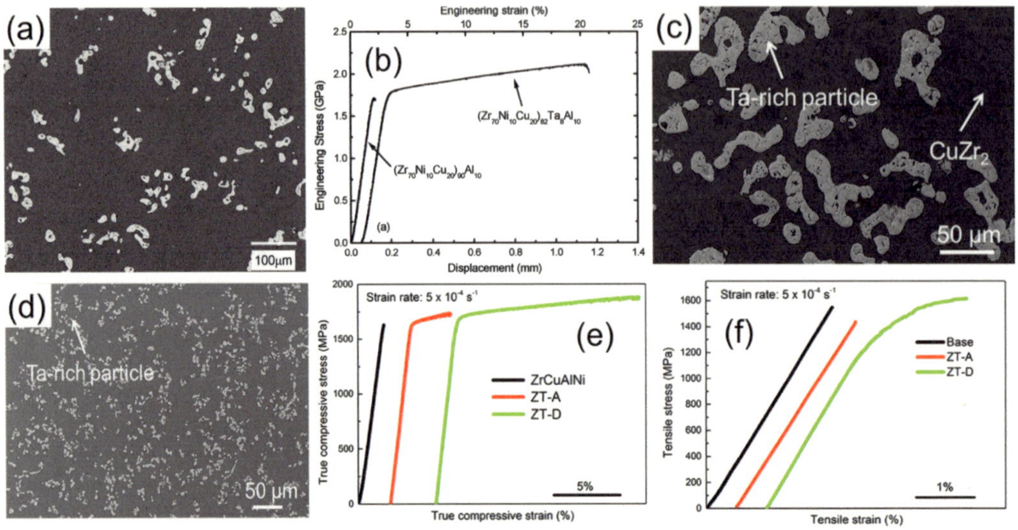

Figure 10. (a) SEM image of in-situ Ta reinforced BMGMCs; (b) compressive stress-strain curves for as-cast in-situ Ta reinforced BMGMCs; (c) SEM images of the sample by conventional arc-melting; (d) SEM images of the sample by novel dealloying method; (e) true compressive stress-strain curves and (f) tensile stress-strain curves of monolithic BMG, BMGMC by arc-melting (ZT-A) and BMGMC by dealloying (ZT-D).

the melt and forming a glass-former liquid, while at the same time, remaining Ta breaks into small particles and distribute in the melt. By finally quenching, the BMGMCs can be fabricated. Interestingly, the size of Ta-rich particles by dealloying method is much smaller than that by conventional arc-melting method, 40 μm of arc-melting sample and 10 μm of dealloying sample, as shown in **Figure 10c** and **d**.

The reason is considered to be that during conventional arc-melting, the Ta-rich particles precipitate during cooling and its size can be hardly controlled, while during dealloying, for the low melting temperature, the Ta-rich particles remain in solid state and finally fine particle can be obtained. With these finer particles, the composite by dealloying method shows better plasticity than that by arc-melting method under both compression and tension, as shown in **Figure 10e** and **f**. The plastic strain for arc-melting sample is 3% under compression and 0% under tension. However, the value for dealloying sample is 14% under compression and 1.8% under tension. As discussed before, the stress concentration at the interfaces between Ta-rich particles and glassy matrix contributes a lot to the overall plasticity. For the finer size of particles in dealloying samples, the number density of particle and equivalent interface area per mm^3 are quite larger than arc-melting sample, 34.4×10^4, 108 mm^2 for dealloying sample and 0.4×10^4 and 22.5 mm^2 for arc-melting sample. Thus, the five times larger area of interfaces in dealloying sample can plasticize the sample more.

Recently, B2-CuZr shape memory secondary phase in Zr-based BMGMCs have attracted a lot of interests. Unlike conventional metal or ceramic reinforcing phases, the shape memory phase can undergo the stress-induced martensitic transformation during deformation, which can both further improve the plasticity and give the sample obvious work hardening behavior [50]. Xu et al. have induced B2-CuZr phase in Zr-Cu-Al-Co system, as shown in **Figure 11a** [51]. The

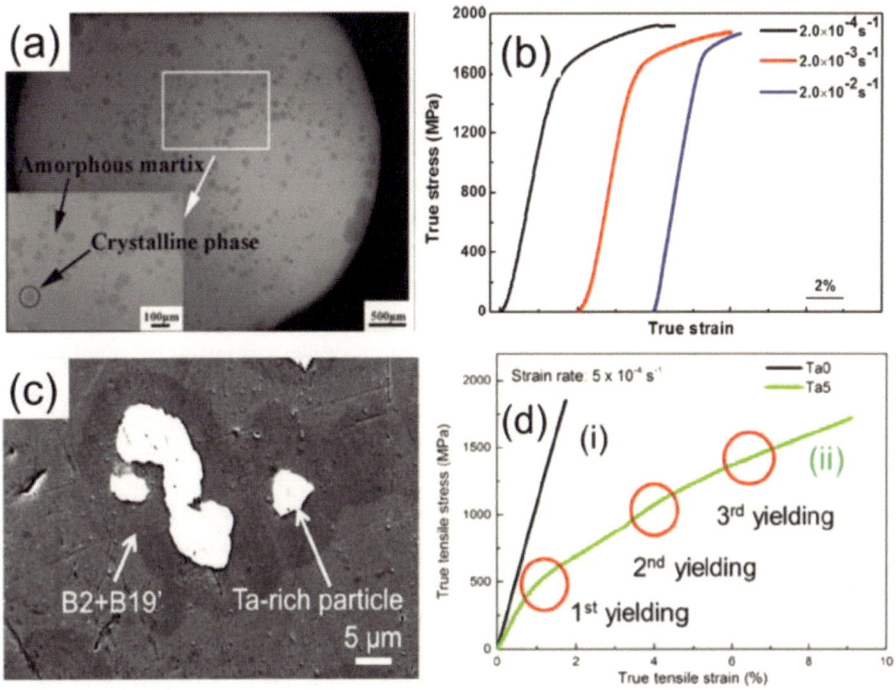

Figure 11. (a) Typical OM images of the B2-CuZr reinforced BMGMCs (inset: enlarged parts of rectangular area); (b) true stress-strain curves of the composites with various strain rates; (c) SEM image of ta-doped sample; (d) true tensile stress-strain curves of both monolithic base alloy (Ta0) and Ta-doped multiphase reinforced BMGMC (Ta5).

spheroidal crystal phases of B2-CuZr (confirmed by XRD, not shown here) are embedded in the amorphous matrix. The volume fraction of B2-CuZr is estimated to be about 10%. The composite exhibits obvious plastic deformation under compression with various strain rates, as shown in **Figure 11b**. Moreover, the flow stress increases with the increasing strain after yielding, exhibiting obvious work-hardening behavior. Even though the B2-CuZr reinforced Zr-based BMGMCs have shown good plasticity and work-hardening, but the inhomogeneous distribution of B2-CuZr limits the further improvement of the mechanical properties. During cooling, the B2-CuZr phase tends to precipitate in the center of the sample where the cooling rate is lower than that near the edge. Guo and Saida have successfully homogenized the distribution of B2-CuZr phase by minor doping Ta [52]. During cooling, the primary precipitated Ta-rich phase acts as effective nucleants that promoted copious nucleation of the B2-CuZr phase. As shown in **Figure 11c**, the dark phase of shape memory phase forms around the gray phase of Ta-rich particle. The total volume fraction of crystalline secondary phase, including both Ta-rich particles and B2-CuZr (some transforms to B19′-CuZr because of the residual heat at the interface), is estimated to be ~80% by comparing the heat of crystallization of both the composites and monolithic BMG. The volume fraction of Ta-rich particles is estimated to be ~10% based on the image analysis. This composite shows a superior plasticity of 8.4% plastic strain during tension, as well as obvious work-hardening and a unique triple yielding phenomenon, as shown in **Figure 11d**.

3.2. Ti-based BMGMCs

The shape memory phase has also been introduced into Ti-based BMGMCs. Hong et al. have successfully induced B2 phase in Ti-Cu-Ni-Zr-Sn-Si system [53]. The size and distribution of B2 phase can be also tailored by varying the composition, as shown in **Figure 12a** and **b**. The size of B2 phase varies from 2–5 to 70–150 μm. The volume fraction varies from 10 to 33%. The composites also shows good plasticity under compression, the largest plastic strain is about 12.7%, as shown in **Figure 12c**. Furthermore, the yield strength exhibits a tendency to decrease with the increase of volume fraction of B2 phase, which originates from the early deformation on softer B2 phase. However, the plastic strain increases with more and larger B2 phase. The large B2 phase is found to be effective in dissipating the localization of shear stress, thus causing branching and multiplication of the shear bands. It is also observable that the severe deformation in the B2 phase, formation of wrinkles suggesting the possible deformation-induced phase transformation, as shown in **Figure 12d**.

Another important reinforcement in Ti-based BMGMCs is the β phase. Very good plasticity can be even observed for such composites under tension [54]. However, most of these composites contains Be, which is toxic and should be avoided when used as biomaterials. To induce β phase in Be-free alloy system, the β phase stabilizers, such as Ta, V and etc. Yamamoto et al. have successfully induced β-Ti phase in Ti-Cu-Ni-Sn-Ta system [55]. As shown in **Figure 13a**, for the low glass-forming ability of the matrix, Ti_2Ni also forms besides β-Ti phase. Thus, the plasticity of this composite is not very good, the plastic strain is about 1.6%, as shown in **Figure 13b**. To further improve the mechanical property, Guo and Kato have chosen a better glass former, Ti-Zr-Cu-Pd-Sn alloy, and used Mo as the β-Ti phase stabilizer element [56]. By doping 2 at%

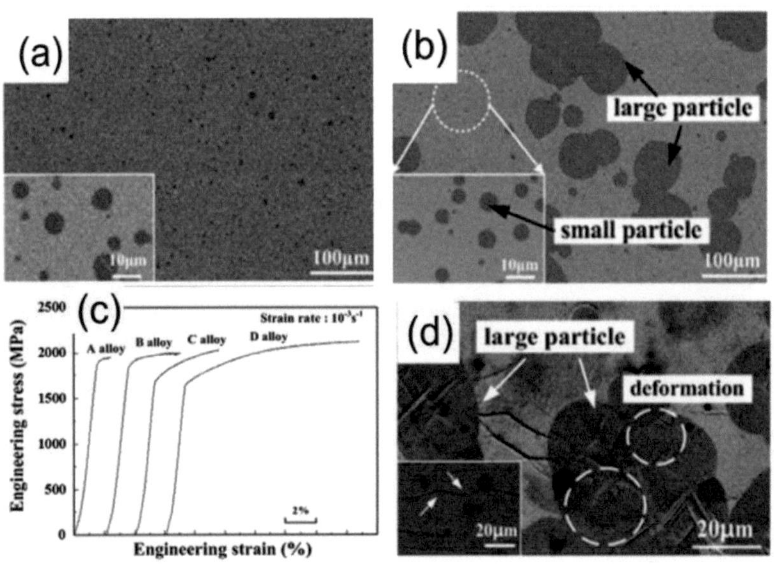

Figure 12. (a) and (b) SEM images of B2 phase reinforced Ti-based BMGMCs with different composition; (c) stress-strain curves of B2 phase reinforced Ti-based BMGMCs; (d) SEM image from the lateral surface of fractured sample.

http://dx.doi.org/10.5772/intechopen.76526

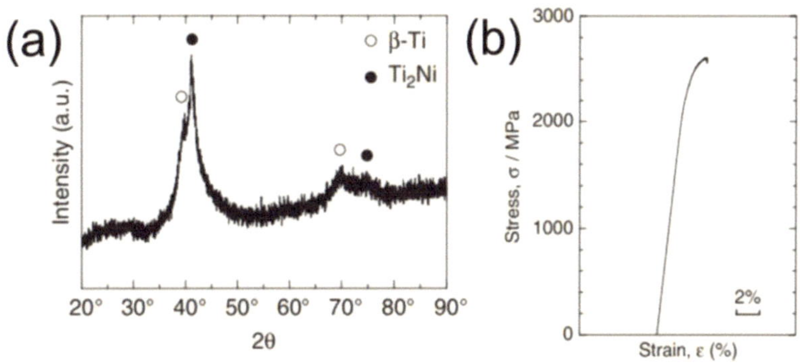

Figure 13. (a) XRD patterns and (b) stress-strain curves under compression for β phase reinforced Ti-Cu-Ni-Sn-Ta BMGMCs.

Mo, homogeneously distributed oblong-like β-Ti phase can be observed from **Figure 14a**. The volume fraction and average size of β-Ti are estimated to be about 25 μm and 25%. This composite also shows a good plasticity under compression, fracture strength of 2160 MPa and plastic strain of 13.4%, as shown in **Figure 14b**. The nanoindentation test has shown that the β-Ti phase is softer than the matrix, indicating the propagation of the main shear band is hindered by the interfaces between the softer β-Ti and glassy matrix. The shear band is deflected, branched, or multiplied. Furthermore, after elastic deformation of both β-Ti and matrix to the yielding point, the β-Ti appears to yield and deform, contributing to the work-hardening behavior. The composite also shows about 3% plastic strain under three point bending test, see **Figure 14c**.

3.3. Mg-based BMGMCs

The researches on in-situ Mg-based BMGMCs are not as much as those in Zr-based or Ti-based systems for the difficulty to design proper fabrication process. However, recently, the application of novel dealloying in metallic melt method or selective phase leaching method in fabrication of in-situ Mg-based BMGMCs have attracted a lot of attentions.

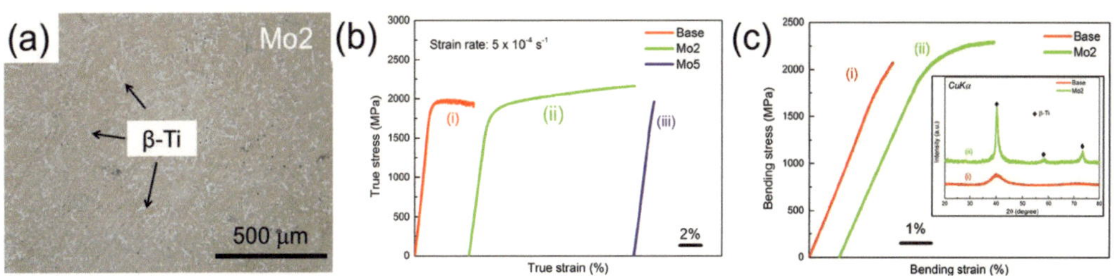

Figure 14. (a) OM image of Mo-doped β-Ti based BMGMCs; (b) compressive true stress-strain curves for both monolithic base alloy and Mo-doped BMGMCs; (c) bending stress-strain curves for both base alloy and BMGMCs (inset: XRD patterns for bending samples).

Oka et al. have successfully introduced α-Ti phase in Mg-Cu-Gd BMG system by using novel dealloying in metallic melt method [57]. The schematic of this method is shown in **Figure 15a**. For the negative value of heat of mixing (miscible) between Ti-Cu and Gd-Cu while positive value (immiscible) of Ti-Gd, Ti-Cu and Gd-Cu phases are expected to form in the prealloy. Similarly, for the negative value of heat of mixing between Mg-Gd and Mg-Cu while positive value of Mg-Ti. When the Ti-Cu-Gd prealloy consisting of Cu-Gd and Ti-Cu phases is immersed in the Mg-melt, the Cu-Gd phase and Cu dealloyed from the Ti-Cu phase are expected to dissolve. They will form the Mg-Cu-Gd BMG formable liquid if the Mg, Cu and Gd proportions are correctly balanced. The remaining elemental Ti from Ti-Cu phase is thought to form the porous structure by a surface diffusion mechanism in the Mg-Cu-Gd alloy liquid. Rapid cooling of the semi-solid mother alloy yields Mg-Cu-Gd BMG with in-situ Ti dispersoids, as shown in **Figure 15b**. By using this strategy, the in-situ Ti dispersoids have been successfully introduced. Furthermore, for the Ti dispersoids are directly dealloyed from Ti-Cu phase, thus, the size of them can be reduced by decreasing the size of Ti-Cu phase. By increasing the cooling rate of Ti-Cu-Gd prealloy, the Ti-Cu phase is refined and subsequently the size of Ti dispersoids also decreased. As shown in **Figure 15c**, during four point bending test, the composite with large and fine Ti dispersoids shows a fracture strength of 230 and 387 MPa, 6 and 78% higher than the monolithic BMG, respectively. With fine pore Ti dispersoids, the size and inter-particle spacing of Ti phase is estimated to be ~500 nm, which is very close to the characteristic plastic processing zone size of reported Mg-based BMG, 100–1000 nm. Thus, optimum condition for the composite effect has locally achieved within and surrounding the porous Ti dispersoids. Therefore, these regions and surrounded area of glassy near porous Ti could deform plastically. However, for the low volume fraction of Ti dispersoids (~2%), macroscopic plasticity is not obtained in this system. Subsequently, Guo et al. have applied similar dealloying reaction in Mg-Cu-Gd-Ag system with better glass-forming ability [49]. As shown in **Figure 16a**, homogenous distributed α-Ti phase among the glassy matrix can be observed. The average size and volume fraction are estimated to be about 6 μm and 13%, respectively. This composite shows improved mechanical properties compared with its monolithic counterpart,

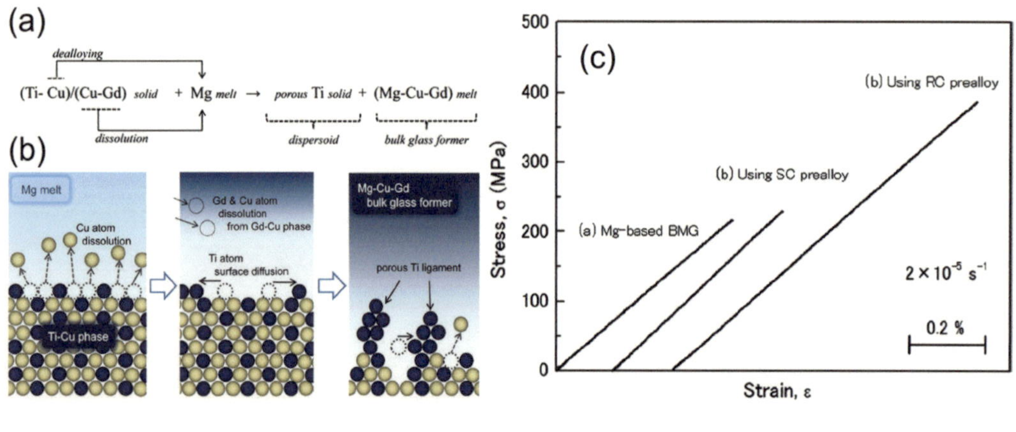

Figure 15. (a) Schematic showing porous Ti formation by dealloying in metallic melt; (b) schematic showing the preparation of the mother alloy; (c) stress-strain curves of BMGMCs using rapid cooling prealloy and slow cooling prealloy, under four point bending test.

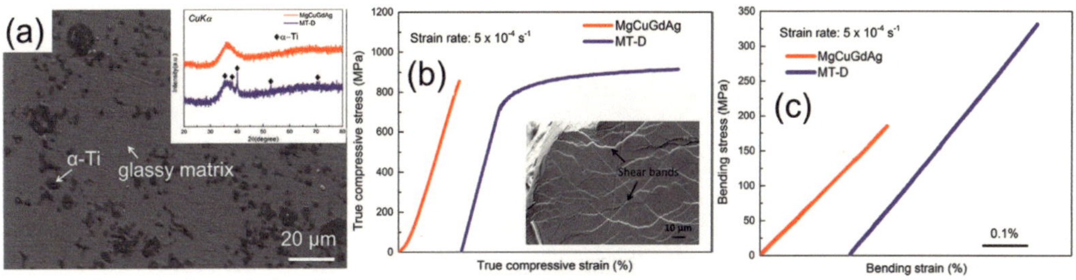

Figure 16. (a) SEM images of in-situ Ti reinforced Mg-based BMGMC by dealloying method (inset: XRD patterns of both BMGMC and its monolithic counterpart); (b) true compressive stress-strain curves of both BMGMC and its base alloy (inset: SEM images of fractured BMGMC); (c) stress-strain curves by four point banding test.

that is, ~6.1% of plastic strain and 920 MPa of fracture strength during compression test, see **Figure 16b**. The stress concentration at the interfaces between Ti dispersoids and surrounding matrix is in favor of initiating multiple shear bands. Furthermore, when the stress exceeds the yield strength of Ti, it can release the stress concentration condition. The suppression of propagation of the single main shear band is enhanced by such yielding, which causes branching, blocking or multiplying the shear bands. It is therefore, the sample is deformed with a significant plasticity. The composite also shows improved fracture stress and strain during four point bending test, 331 MPa of fracture stress and 0.42% of fracture strain, as shown in **Figure 16c**. The calculated fracture toughness of this composite is ~1.73 MPa $m^{1/2}$, 45% higher than its monolithic counterpart. However, it is still low and cannot lead to plasticity during bending.

As stated above, the shape memory phase has attracted a lot of interest recently for its unique stress-induced martensitic transformation behavior. However, the research on shape memory phase reinforced BMGMCs mainly focus on Zr-based or Ti-based BMGMCs. Guo and Kato have successfully induced in-situ B2-NiTi shape memory phase in Mg-Ni-Gd-Ag BMGMCs by using novel selective phase leaching in metallic melt method [58]. A schematic of the novel designed process is shown in **Figure 17a**, which contains roughly three steps: Ni-Ti-Gd precursor preparation by arc-melting, master alloy preparation by induction melting and composite preparation by copper mold casting. From the Ni-Ti, Ti-Gd and Ni-Gd phase diagrams, it is possible to prepare a Ni-Ti-Gd ternary precursor consisting of only NiTi and Ni-Gd phases if the proportions of Ti, Ni, and Gd are properly balanced. Then, the temperature of master alloy preparation is kept low enough for NiTi dispersoids not to melt or dissolve, but high enough for the Ni-Gd phase to dissolve into the Mg-Ag melt, owing to their different reactivities. Moreover, the amounts of Mg-Ag melt and dissolved Ni-Gd was properly balanced to form the glass-forming matrix. Finally, Mg-based BMGMCs, an Mg-Ni-Gd-Ag BMG matrix with in-situ NiTi dispersoids, is fabricated by casting this semi-solid melt into a copper mold. By using such strategy, the in-situ B2-NiTi phase have been successfully induced, as shown in **Figure 17b**, the average size is ~8 μm and the volume fraction is ~15%. As shown in **Figure 17c**, the composite shows a higher fracture stress (~906 MPa), plastic strain (~7%), and work hardening than its monolithic counterpart. The shear bands were considered to be obstructed by the in-situ ductile NiTi dispersoids, which deflected their propagation and caused branching or multiplying of the shear bands, typically observed in ductile metal reinforced BMGMCs and is known as the "blocking effect". Moreover, a stress-induced phase transformation from

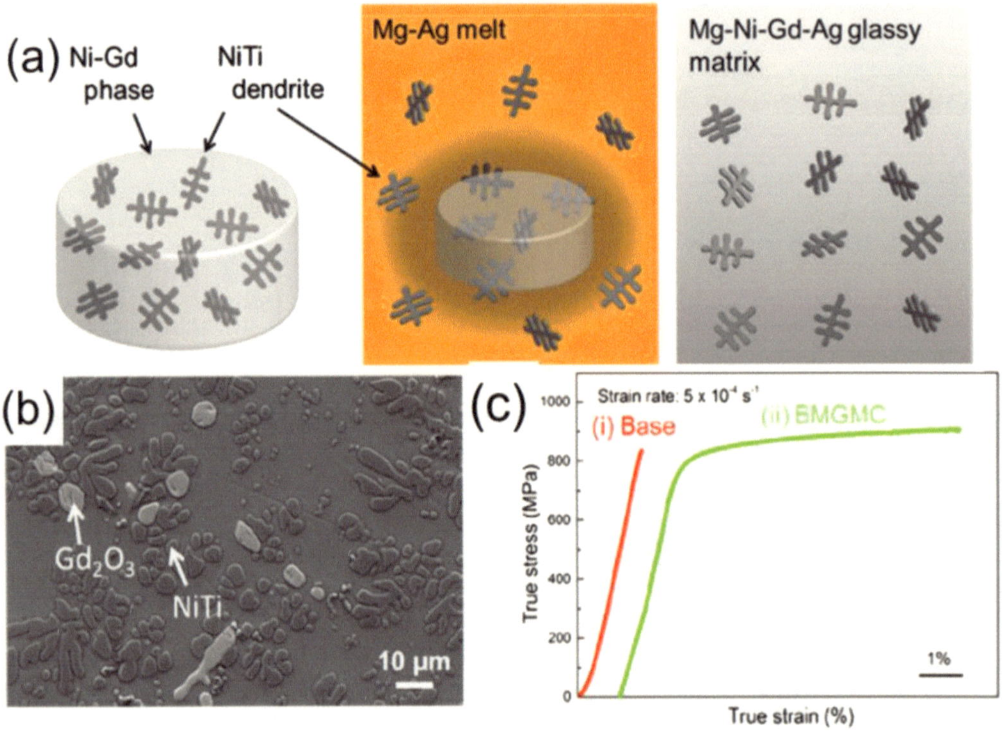

Figure 17. (a) Schematic of the fabrication process; (b) SEM image of B2-NiTi reinforced mg-based BMGMC; (c) compressive true stress-strain curves for both base alloy and BMGMC.

B2-NiTi to B19′-NiTi during deformation releases the stress concentration around the dispersoids to restrict free volume accumulation. This process hindered the rapid propagation of shear bands, meaning that additional stress was required to move the shear bands, called the "TRIP effect". Since the B2-NiTi phase precipitates during Ni-Ti-Gd precursor preparation and has no change from then on, the size of it can be refined by increasing the cooling rate of the precursor. Thus, the precursor rods by tilt casting technique are produced with various diameters, i.e., various cooling rate. With a higher cooling rate, the size of B2-NiTi is smaller, from 8 to 2 μm, as shown in **Figure 18a** and **b**. With finer B2-NiTi phase, the composite shows higher fracture strength and larger plastic strain, as shown in **Figure 18c**. The sample with finest particle size exhibits the best mechanical properties, i.e., 1096 MPa fracture strength and 15.5% plastic strain. The volume fraction of the B2-NiTi phase has also been improved by adjusting the composition with higher Ti amount. The volume fraction increases from 15 to 32% and the optimized composite shows superior plasticity during compression, as shown in **Figure 19a**, a fracture stress of 1212 MPa and a fracture strain of 25.3%. **Figure 19b** summarizes the compressive property data of various in-situ Mg-based BMGMCs, including Fe, long-period stacking ordered structure (LPSO), NiZr, AgMg, and quasicrystal reinforced composites [59–63]. Both the fracture strength and fracture strain of the optimized B2-NiTi reinforced BMGMC are the highest among all in-situ Mg-based BMGMCs reported to date.

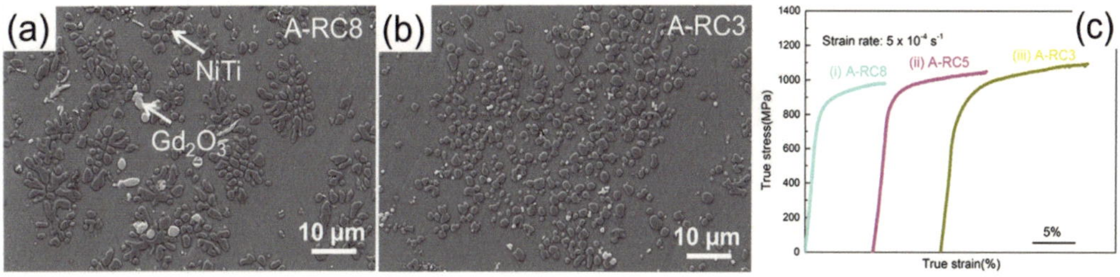

Figure 18. SEM images of the composites using rapid cooling precursor: (a) 8 mm rod precursor; (b) 3 mm rod precursor; (c) compressive true stress-strain curves of the composites using rapid cooling precursor with various diameter, 8, 5 and 3 mm.

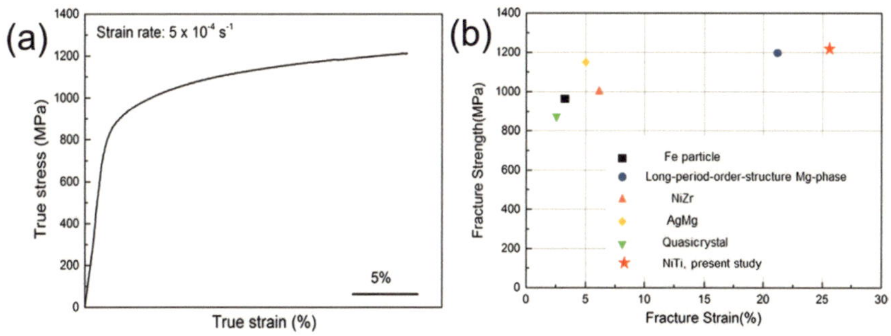

Figure 19. Compressive true stress-strain curves of the optimized B2-NiTi reinforced BMGMCs; (b) fracture strength versus fracture strain for various in-situ Mg-based BMGMCs to date.

4. Conclusion and outlook

In this chapter, both ex-situ and in-situ BMGMCs developed in Zr-based, Ti-based, Mg-based systems have been introduced, such as ceramic particle, metal particle, porous particle reinforced ex-situ BMGMCs and B2-phase, β-phase reinforced in-situ BMGMCs. The microstructures, mechanical properties as well as deformation mechanisms are discussed for each kind of BMGMCs. Compared with nearly zero plasticity of monolithic BMGs, the BMGMCs reinforced by secondary phases show significant improvement in plasticity, e.g., β-phase reinforced Ti-based BMGMCs show over 10% plastic strain under tension, B2-NiTi reinforced Mg-based BMGMCs show over 20% plastic strain under compression, etc. For the limitation of the chapter, more works on various reinforcements and alloy systems cannot be covered. The detailed deformation mechanisms of BMGMCs are not discussed fully either. For further development of BMGMCs, more works should be done on more complex composite structure, the deformation mechanisms, designing novel processing methods, tailoring the microstructures and mechanical properties of the existed BMGMCs. The research on BMGMCs will greatly extend the application potentials of amorphous materials as engineering materials.

Author details

Wei Guo

Address all correspondence to: weiguo@hust.edu.cn

State Key Lab of Materials Processing and Die and Mould Technology, Huazhong University of Science and Technology, Wuhan, China

References

[1] Klement T, Willens R, Duwez P. Non-crystalline structure in solidified gold–silicon alloys. Nature. 1960;**187**:869

[2] Conner RD, Dandliker RB, Johonson WL. Mechanical properties of tungsten and steel fiber reinforced $Zr_{41.25}Ti1_{3.75}Cu12.5Ni_{10}Be_{22.5}$ metallic glass matrix composites. Acta Materialia. 1998;**46**:455

[3] Peker A, Johonson WL. A highly processable metallic glass: $Zr_{41.2}Ti_{13.8}Cu_{12.5}Ni_{10.0}Be_{22.5}$. Applied Physics Letters. 1993;**63**:2342

[4] Inoue A, Fan C, Saida J, Zhang T. High-strength Zr-based bulk amorphous alloys containing nanocrystalline and nanoquasicrystalline particles. Science and Technology of Advanced Materials. 2000;**1**:73

[5] Drehman A, Greer A, Turnbull D. Bulk formation of a metallic glass: $Pd_{40}Ni_{40}P_{20}$. Applied Physics Letters. 1982;**41**:716

[6] Kui H, Greer A, Turnbull D. Formation of bulk metallic glass by fluxing. Applied Physics Letters. 1984;**45**:615

[7] Turnbull D. The liquid state and the liquid-solid transition. Transactions of AIME. 1961;**221**:422

[8] Telford M. The case for bulk metallic glass. Materials Today. Materials Today. 2004:36

[9] Liu C, Heatherly L, Easton D, Carmichael C, Schnerberl J, Chen C. Test environments and mechanical properties of Zr-base bulk amorphous alloys. Metallurgical and Materials Transactions A: Physical Metallurgy and Materials Science. 1998;**29**:1811

[10] Inoue A. Stabilization of metallic supercooled liquid and bulk amorphous alloys. Acta Materialia. 2000;**48**:279

[11] Inoue A, Zhang W, Zhang T, Kurosaka K. Cu-based bulk glassy alloys with high tensile strength of over 2000 MPa. Journal of Non-Crystalline Solids. 2002;**304**:200

[12] Inoue A, Shen BL, Koshiba H, Kato H, Yavari AR. Ultra-high strength above 5000 MPa and soft magnetic properties of Co-Fe-Ta-B bulk glassy alloys. Acta Materialia. 2004;**52**:1631

[13] Inoue A. High Strength Bulk Amorphous Alloys with Low Critical Cooling Rates (Overview). Materials Transactions, JIM. 1995;**36**:866

[14] Schroeder V, Gilbert CJ, Ritchie RO. Comparison of the Corrosion Behavior of a Bulk Amorphous Metal, $Zr_{41.2}Ti_{13.8}Cu_{12.5}$Ni10Be22.5, with Its Crystallized Form Scripta Materialia. 1998;**38**:1481

[15] Chen H, He Y, Shiflet GJ, Poon SJ. Mechanical properties of partially crystallized aluminum based metallic glasses. Scripta Materialia. 1991;**25**:1421

[16] Argon AS, Kuo HY. Plastic flow in a disordered bubble raft (an analog of a metallic glass). Materials Science and Engineering. 1979;**39**:101

[17] Falk ML, Langer JS. Dynamics of viscoplastic deformation in amorphous solids. Physical Review E. 1998;**57**:7192

[18] Lemaitre A. Rearrangements and dilatancy for sheared dense materials. Physical Review Letters. 2002;**89**:195503

[19] Lund AC, Schuh CA. Yield surface of a simulated metallic glass. Acta Materialia. 2003;**51**:5399

[20] Lund AC, Schuh CA. The Mohr–Coulomb criterion from unit shear processes in metallic glass. Intermetallics. 2004;**12**:1159

[21] Johnson L, Samwer K. A universal criterion for plastic yielding of metallic glasses with a (T/T-g)(2/3) temperature dependence. Physical Review Letters. 2005;**95**:195501

[22] Mayr SG. Activation energy of shear transformation zones: A key for understanding rheology of glasses and liquids. Physical Review Letters. 2006;**97**:195501

[23] Zink M, Samwer K, Johnson WL, Mayr SG. Plastic deformation of metallic glasses: Size of shear transformation zones from molecular dynamics simulations. Physical Review B. 2006;**73**:172203

[24] Chen LY, Setyawan AD, Kato H, Inoue A, Zhang GQ, Saida J, Wang XD, Cao QP, Jiang JZ. Free-volume-induced enhancement of plasticity in a monolithic bulk metallic glass at room temperature. Scripta Materialia. 2008;**59**:75

[25] Schuh CA, Hufnagel TC, Ramamurty U. Mechanical behavior of amorphous alloys. Acta Materialia. 2007;**55**:4067

[26] Joo SH, Kato H, Kim HS. Work-Hardening Induced Tensile Ductility of Bulk Metallic Glasses via High-Pressure Torsion. Scientific Reports. 2015;**5**:9660

[27] Chu JP, Greene JE, Jang JSC, Huang JC, Shen YL, Liaw PK, Yokoyama Y, Inoue A, Nie TG. Bendable bulk metallic glass: Effects of a thin, adhesive, strong, and ductile coating. Acta Materialia. 2012;**60**:3226

[28] Dandliker RB, Conner RD, Johnson WL. Melt infiltration casting of bulk metallic-glass matrix composites. Journal of Materials Research. 1998;**13**:2896

[29] Conner RD, Dandliker RB, Johnson WL. Mechanical properties of tungsten and steel fiber reinforced $Zr_{41.25}Ti_{13.75}Cu_{12.5}Ni_{10}Be_{22.5}$ metallic glass matrix composites. Acta Materialia. 1998;**46**:6089

[30] Kim CP, Busch R, Masuhr A, Choi-Yim H, Johnson WL. Processing of carbon-fiber-reinforced $Zr_{41.2}Ti_{13.8}Cu_{12.5}Ni_{10.0}Be_{22.5}$ bulk metallic glass composites. Applied Physics Letters. 2001;**79**:1456

[31] Qiu KQ, Suo ZY, Ren YL, Yu B. Observation of shear bands formation on tungsten fiber-reinforced Zr-based bulk metallic glass matrix composite. Journal of Materials Research. 2007;**22**:551

[32] Li JC, Chen XW, Huang FL. FEM analysis on the deformation and failure of fiber reinforced metallic glass matrix composite. Materials Science and Engineering A. 2016;**652**:145

[33] Zhang H, Zhang ZF, Wang ZG, Zhang HF. Deformation and damage evolution of tungsten fiber reinforced metallic glass matrix composite induced by compression. Materials Science and Engineering A. 2008;**483-484**:164

[34] Qiu KQ, Wang AW, Zhang HF, Ding BZ, Hu ZQ. Mechanical properties of tungsten fiber reinforced ZrAlNiCuSi metallic glass matrix composite. Intermetallics. 2002;**10**:1283

[35] Choi-Yim H, Johnson WL. Bulk metallic glass matrix composites. Applied Physics Letters. 1997;**71**:3808

[36] Zhang QS, Zhang W, Xie GQ, Inoue A. Unusual Plasticity of the Particulate-Reinforced Cu-Zr-Based bulk metallic glass composites. Materials Transactions. 2007;**48**:2542

[37] Conner RD, Choi-Yim H, Johnson WL. Mechanical properties of $Zr_{57}Nb_{5}Al_{10}Cu_{15.4}Ni_{12.6}$ metallic glass matrix particulate composites. Journal of Materials Research. 1999;**14**:3292

[38] Choi-Yim H, Conner RD, Szuecs F, Johnson WL. Processing, microstructure and properties of ductile metal particulate reinforced $Zr_{57}Nb_{5}Al_{10}Cu_{15.4}Ni_{12.6}$ bulk metallic glass composites Acta Materialia. 2002;**50**:2737

[39] Xu YK, Ma H, Xu J, Ma E. Mg-based bulk metallic glass composites with plasticity and gigapascal strength Acta Materialia. 2005;**53**:1857

[40] Liu T, Shen P, Qiu F, Zhang T, Jiang Q. Advanced Engineering Materials. 2009;**11**:392

[41] Chen C, Xue Y, Wang L, Cheng X, Wang F, Wang Z, Zhang H, Wang A. Microstructures and Mechanical Properties of ZrC Reinforced (Zr-Ti)-Al-Ni-Cu Glassy Composites by an In Situ Reaction. Advanced Engineering Materials. 2012;**14**:439

[42] Xue YF, Cai HN, Wang L, Wang FC, Zhang HF. Strength-improved Zr-based metallic glass/porous tungsten phase composite by hydrostatic extrusion. Applied Physics Letters. 2007;**90**:081901

[43] Pan DG, Zhang HF, Wang AM, Hu ZQ. Enhanced plasticity in Mg-based bulk metallic glass composite reinforced with ductile Nb particles. Applied Physics Letters. 2006;**89**:261904

[44] Wada T, Inoue A, Greer AL. Enhancement of room-temperature plasticity in a bulk metallic glass by finely dispersed porosity. Applied Physics Letters. 2005;**86**:251907

[45] Jang JSC, Ciou JY, Hung TH, Huang JC, Du XH. Enhanced mechanical performance of Mg metallic glass with porous Mo particles. Applied Physics Letters. 2008;**92**:011930

[46] Guo W, Wada T, Kato H. Work-hardenable Mg-based bulk metallic glass matrix composites reinforced by ex-situ porous shape-memory-alloy particles. Materials Letters. 2016;**183**:454

[47] Hays CC, Kim CP, Johnson WL. Microstructure controlled shear band pattern formation and enhanced plasticity of bulk metallic glasses containing in situ formed ductile phase dendrite dispersions. Physical Review Letters. 2000;**84**:2901

[48] Fan C, Ott RT, Hufnagel TC. Metallic glass matrix composite with precipitated ductile reinforcement. Applied Physics Letters. 2002;**81**:1020

[49] Guo W, Kato H, Yamada R, Saida J. Fabrication and mechanical properties of bulk metallic glass matrix composites by in-situ dealloying method. Journal of Alloys and Compounds. 2017;**707**:332

[50] Hofmann DC. Shape memory bulk metallic glass composites. Science. 2010;**329**:1294

[51] Xu J, Ma L, Xue Y, Nie Z, Long Y, Wang L, Deng Y. Work-hardening behavior, strain rate sensitivity, and failure behavior of in situ CuZr-based metallic glass matrix composite. Journal of Materials Science. 2016;**51**:5992

[52] Guo W, Saida J. Triple-yieldable multiphase reinforced bulk metallic glass matrix composites under tension. Materials Letters. 2017;**191**:42

[53] Hong SH, Kim JT, Lee MW, Park JM, Lee MH, Kim BS, Park JY, Seo Y, Suh JY, Yu P, Qian M, Kim KB. Combinatorial Influence of Bimodal Size of B2 TiCu Compounds on Plasticity of Ti-Cu-Ni-Zr-Sn-Si Bulk Metallic Glass Composites. Metallurgical and Materials Transactions A: Physical Metallurgy and Materials Science. 2014;**45A**:2376

[54] Hofmann DC, Suh JY, Wiest A, Lind ML, Demetriou MD, Johnson WL. Development of tough, low-density titanium-based bulk metallic glass matrix composites with tensile ductility. Proceedings of the National Academy of Sciences of the United States of America. 2008;**105**:20136

[55] Yamamoto T, Ito H, Hasegawa M, Inoue A. Mechanical properties and microstructures of composites of Ti-based metallic glass and β-Ti. Materials Transactions. 2007;**48**:1812

[56] Guo W, Kato H. Development of in-situ β-Ti reinforced Be-free Ti–based bulk metallic glass matrix composites. Journal of Alloys and Compounds. 2017;**714**:120

[57] Oka H, Guo W, Wada T, Kato H. Mg-based metallic glass matrix composite with in situ porous titanium dispersoids by dealloying in metallic melt. Materials Science and Engineering A. 2013;**582**:76

[58] Guo W, Kato H. Development and microstructure optimization of Mg-based metallic glass matrix composites with in situ B2-NiTi dispersoids. Materials and Design. 2015;**83**:238

[59] Ma H, Xu J, Ma E. Mg-based bulk metallic glass composites with plasticity and high strength. Applied Physics Letters. 2003;**83**:2793

[60] Hui X, Dong W, Chen GL, Yao KF. Formation, microstructure and properties of long-period order structure reinforced Mg-based bulk metallic glass composites. Acta Materialia. 2007;**55**:907

[61] Chen G, Zhang XL, Liu CT. High strength and plastic strain of Mg-based bulk metallic glass composite containing in situ formed intermetallic phases. Scripta Materialia. 2013;**68**:150

[62] Wang SG, Xu J. Strengthening and toughening of Mg-based bulk metallic glass via in-situ formed B2-type AgMg phase. Journal of Non-Crystalline Solids. 2013;**379**:40

[63] Zhao YY, Men H, Estévez D, Liu Y, Wang XM, Li RW, Chang CT. Mg-based bulk metallic glass composite containing in situ microsized quasicrystalline particles. Scripta Materialia. 2014;**78-79**:21

Section 5

Metallic Glasses for Tribo-electrochemistry Systems

Chapter 5

Metallic Glasses for Triboelectrochemistry Systems

Abdenacer Berradja

Additional information is available at the end of the chapter

http://dx.doi.org/10.5772/intechopen.78233

Abstract

A fundamental quest of modern triboelectrochemistry is to unravel the prevailing failure mechanisms when surface interactions are operated in corrosive environments and to study how these influence the performance of materials and tools. Both system and materials oriented approaches are thus required to deal with the electrochemical and physico-chemical changes of matter due to the influence of a mechanical sliding energy between the two contacting surfaces (i.e. tribocorrosion damage). In this chapter, metallic glass material concepts used in environments where tribocorrosion occurs are described. Concepts to act in opposition to wear and corrosion are briefly reported. In particular, a description is given of different groups of metallic glasses designed to withstand the effects of combined and uncoupled wear and corrosion. Metallic glass composition, structural effect, crystallization level, passive film formation, microhardness are often viewed as the most critical elements in the performance of tribocorrosion of metallic glasses. Interactions between wear and corrosion mechanisms are identified along with some models that aim to inform metallic glass selection and predict performance.

Keywords: wear-resistant bulk metallic glasses, corrosion-resistant bulk metallic glasses, wear-corrosion synergy, tribocorrosion, triboelectrochemistry

1. Introduction

Metals and their alloys of quasi-crystalline structures have been widely used for thousands of years, commencing with the Bronze Age, which took place approximately 3000 down to 100 years, and passing through the Iron Age [1]. Although, the utilization of these common engineering materials, namely ferrous and nonferrous, are still in current use nowadays, however, they are experiencing much concurrence with a range of novel materials and a bewildering array of solid products due to the engineering progress, which essentially depends on the availability and the intelligent use of materials.

At the present time, there are available for use in excess of 45,000 different metallic alloys [1]. Albeit, the steels and cast irons make up the largest use on a weight basis, the number of different nonferrous alloys exceed the number of ferrous alloys. The primary nonferrous alloys are those in which the base metal consists of either aluminum, copper, nickel, magnesium, titanium, zirconium or zinc [1].

With the introduction of new metallic alloys and the breakthrough in the production of the so-called glassy metals, what was the best choice several years ago may no longer be so. Over the years, considerable efforts and great progress have been made in the field of materials selection, in particular through the improvement of the specific properties of different sorts of alloys. These growing developments, inter alia, include processes for enhancing their metallurgical, mechanical, physical, chemical, and especially their tribocorrosion properties. Alternatives in the composition have also been formulated to improve the workability (e.g., glass-forming ability) of many metallic glass alloys through, commonly, monitoring the quenching rate. Critical cooling rate and maximum attainable size, known as critical casting thickness, are both direct indicators of glass-forming ability (GFA). The smaller the critical cooling rate and/or the larger maximum attainable size, the higher is the GFA.

In order to conduct a meaningful evaluation of a design alloy, all essential data required to fit with the most appropriate material must be disposable. It is the purpose of this chapter to supply as much of this information as possible for commercially available metallic glass materials, and for use in systems where mechanical and chemical surface interactions take place and leading to deterioration effects. The main strategies known to meet these requirements are outlined. Since mechanical wear and corrosion are broad fields, where the interplay between several mechanisms can occur, different approaches to prevent them have been contracted. Some of the most common ways are discussed. Examples from different classes of metallic glass material are given.

2. A short overview of mechanical properties of metallic glasses

A close relationship exists between properties of materials whose magnitude is usually determined by their bond strength. The choice of a material is the result of several compromises. For instance, the technical appraisal of an alloy will generally be a compromise between mechanical resistance, such as fracture toughness and some other properties such as yield strength, Young's modulus, density etc. A convenient way of bringing out these relationships is by a series of figures or charts, in which one parameter is plotted as a function of another.

An important relationship is that between fracture toughness and the yield strength, a typical characteristic of the damage tolerance of materials. Values of most engineering materials, including monolithic metallic glasses, and ductile-phase-reinforced metallic glass composites [2] are shown in **Figure 1**.

Generally, ductile metals have very high fracture toughness and fairly low yield strength. Metallic glasses, however, show limited plastic yielding and have toughness-strength (K_c-σ_y) relationships that lie between brittle ceramics and marginally tough materials. The lack of

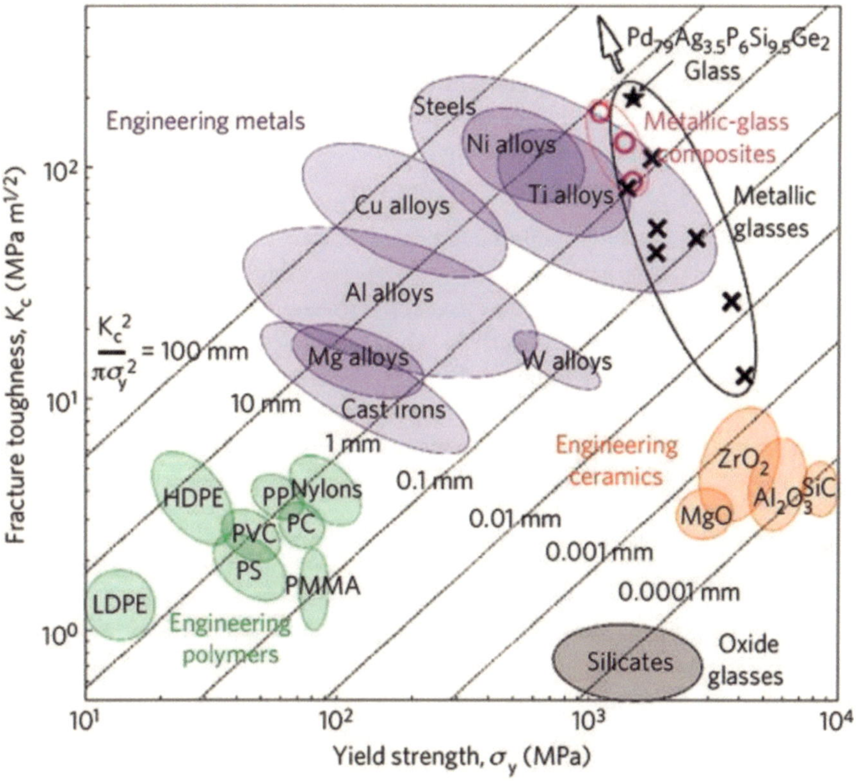

Figure 1. Ashby map of the damage tolerance of materials (fracture toughness *vs.* yield strength) including oxide glasses, ceramics, polymers, metals, monolithic metallic glasses (Fe-based glasses, Zr-based glasses, Ti-based glass, Pt-based glass) designated by crosses, Pd-based glass designated by a filled star, and ductile-phase-reinforced metallic glass composites designated by circles. (reproduced from Demetriou et al. [2] with permission from copyrighted ©Nature, 2011).

plasticity hinders the application of bulk metallic glasses (BMGs) as structural materials, and prevent their usage for instance in load-bearing structures. The need for the development of new efficient and resistant materials is thus essential in order to ensure this objective.

In this regard, great improvements in fracture toughness have been made as new alloy compositions were discovered and optimized to extend the fracture toughness limit for BMGs. The occupation of a new region of property space for BMG composites (BMGCs), designated by circles, is shown in the Ashby's map (**Figure 1**) [2]. Contours correspond to values for the plastic-zone radius, $K^2_c/\pi\sigma^2_y$. In this zone area, plastic deformation predominantly occurs at the crack tip because of the high stresses generated by the sharp stress concentration [3]. It is mostly this enhanced plasticity scale in BMGCs, which is at the origin of their superiority over other technical candidate materials. It is argued to be due to the fact that all the shear bands initiated in plastically soft regions with lower yield stress (or lower shear modulus) are prevented in the surrounding regions of higher yield stress or stiffness. This enhancement in both ductility and toughness is analogous to the tempering of plastics by inclusion of rubber particles [4]. As indicated by the arrow, the conjunction of toughness to the strength, potentially accessible to metallic glasses extends beyond traditional benchmarks towards levels formerly inaccessible to any material (e.g., Pd-based alloys in [2]). One direct result of the unique microstructure is the high toughness-to-yield strength ratio, mainly accessible to BMGCs. Their strength exceeds that of the strength limit of known crystalline pure metals or alloys and approaches that of engineering ceramics, whereas their toughness is markedly high, among metallic alloys.

Some recent significant developments have been made towards the design of this kind of BMGC materials. This was achieved through the successful implementation of effective composite microstructures, which typically combine a strong glassy matrix with ductile crystalline reinforcements that suppress fracture while sustaining high strength. This variety of composite materials can only be obtained through the commitment of a nanocrystallization process [5] or *via* the reinforcement with ceramic particles [6]. Current studies on BMGCs performance are still at the development stage and concern the evaluation of either their mechanical properties or their corrosion resistance, but the perspective is very promising.

In recent years, a wide variety of industries including food, medical and pharmaceutical, aircraft components, electronics, building materials, and automobile industries have been promoting the technological development of newly composite materials including the vitreous-based composites to achieve suitable strength/density, and toughness/stiffness ratios.

3. Tribocorrosion fundamentals

An important problem in tribology concerns the interaction between friction processes and electrochemical reactions (corrosion) occurring in reactive environments such as aqueous media or hot aggressive gases. The effect of mechanical stimuli on chemical degradation of materials and, *vice-versa,* the influence of corrosion on the mechanical response of contacting materials are of great concern in many areas of tribology. A new research area, "tribocorrosion", emerged in recent years, mainly driven by the increasing demand from the biomedical implant, power generation, marine, and offshore industries. Since tribocorrosion is ubiquitous in many technical applications, it is necessary to employ material means capable of withstanding its damage effects.

Tribocorrosion can be defined as "…the process leading to a material degradation (i.e., material loss) which results from simultaneous mechanical (materials properties, surface and sub-surface transformations, cracking, etc.), chemical, electrochemical (corrosion attack) and/or biological material removal mechanisms (i.e., bio-tribocorrosion)…" [7]. It is basically an integration of two major areas of significance and application in mechanical systems, namely Tribology and Corrosion [8, 9]:

Tribocorrosion involves the inter-play between friction, wear (tribological), and corrosion (electrochemical) phenomena in a complex way. This includes such diverse phenomena as wear-accelerated corrosion, fretting-corrosion, erosion-corrosion, oxidational wear, chemical or corrosive wear often described in the literature. Therefore, tribocorrosion (or tribological) processes are intricate and not intrinsically fundamental, in the sense that Young's modulus for example is fundamental, but rather that they depend on the triboelectrochemical system approach. In particular they are determined by a combination of a number of more fundamental properties of the contacting materials, testing parameters, and test conditions, especially the nature of the environment in which the tests take place. An important aim of research in tribocorrosion field is precisely that of determining the nature of this dependence, so that the triboelectrochemical behavior may be predicted from a knowledge of a system approach and the more fundamental properties of interacting surfaces. Although, this aim has not been achieved yet, a fair progress has been made, and still more work is required to have a good comprehension of just which are the important distinctive features determining the surface interaction behavior in a triboelectrochemical system.

3.1. Elements of tribocorrosion (instrumentation)

Tribocorrosion of two contacting solids in relative motion is, just as friction, a system parameter. A triboelectrochemical system consists of implementing electrochemical techniques to a tribological designed system (i.e., tribometer type, and complete material system). That mechanical designed system is of a great importance since this will enable to simulate as much as possible the entire material system used in the field, and the constraints that have associated with it (e.g., similarity of the wear mechanisms active in the laboratory test and in the field, such as abrasion, adhesion, fatigue, penetration hardness, bending, existence or not of a third body, erosion, corrosion, their combinations, etc.).

Generally, the tribological configuration involves:

- The choice of a tribometer type with respect to its characteristics (e.g., contact noise, vibrations, residual stress …);
- The choice of a body (material specimen under test), and a counter-body (material antagonist, usually a ball, a pin, a plate or a disk), which constitutes the contact system geometry (e.g., sphere-on-flat, sphere-on-sphere, cylinder-on-flat, flat-on-flat…). This choice has to consider the metallurgical and chemical features of the two solid bodies (e.g., inertness, composition of contacting materials, microstructure, surface film composition, etc.);
- The choice of a lubricant or other matter between the two bodies (e.g., composition of the medium, viscosity, solid particles in suspension, stagnant or stirred, etc.);
- The choice of surrounding media and temperature (e.g., relative humidity, vacuum, composition of the corrosive environment, pH, aggressiveness, ionic conductivity, etc.);
- The loading conditions (pressure or force, relative velocity, acceleration, or frequency, a working distance, or number of contacts);
- And finally, the choice of the type of relative movement (uni- or bidirectional, continuous or reciprocating).

The measuring instruments in tribocorrosion tests allow to monitor in real-time and on-line the foregoing system parameters (e.g., contact conditions: normal or tangential force, relative displacement, velocity, etc.). The main focus is to promptly control any variation or change associated with these parameters or settings *in-situ*. They need also to be instrumented with electrochemical techniques, which enables the management and the recording of applied electrochemical parameters and/or their responses (e.g., polarization of the contacting materials, charge density, etc.). The ultimate goal is to obtain promptly any information on the evolution of the input and output chemical-mechanical measurements during the test. These *in-situ* data outcome combined with *ex-situ* surface characterization techniques (e.g., high resolution SEM imaging, TEM, XRD, EDAX, FIB, XPS, Auger spectroscopy, FT-IR, roughness surface profilometry, micro- or nanoindentation hardness, etc.) and chemical analyses of post-test solutions (e.g., ICP-AES, ICP-MS) allow for the disclosure of the wear-corrosion mode, and thereby contributing to a better understanding of the tribocorrosion mechanisms involved (corrosion mechanism, wear regime, friction process, etc.).

The typical configuration of a triboelectrochemical cell experiment involves an inert material (e.g., corundum counter-body) sliding against the investigated material (i.e., working electrode) under mechanoelectrochemical well-controlled conditions. Although metal-on-metal contact configurations are possible, but the use of an inert counter-body simplifies the interpretation of the electrochemical results of the corrosion-wear process. In addition, it is most advising to perform the electrochemical measurements under stationary regime conditions, at least prior to starting up of the measurements. The development of relevant models for the interpretation of the tribocorrosion mechanism concurrently depends, essentially, on the choice of electrochemical techniques to be implemented in a tribocorrosion test and the mechanical contact conditions (e.g., relative motion). The state-of-the art and reviews on triboelectrochemical techniques and experiments are available elsewhere [8–13].

3.2. The use of electrochemical techniques in the study of *in-situ* tribocorrosion processes

Corrosion processes are of electrochemical nature and therefore electrochemistry likely interferes with the tribological behavior of tribocorrosion systems. Attempts were made in recent years to control wear by electrochemical methods in engineering and biomedical systems. These electrochemical techniques provide a very convenient way to measure the rate of corrosion processes in either the laboratory or the field. Such methods can also be used in many different ways to assess either their efficiency of monitoring corrosion material degradation (e.g., concrete steels, and marine alloy structures) or their capability for material protection (e.g., inhibitors, protective layers, coatings, and appropriate metals and alloys). The theoretical bases and the practical implementation of the electrochemical techniques have been published in books and review articles [8, 9, 11, 14] and this material is compendiously repeated here only to the extent that it is needed to define the terminology as well as the utility of the topic for the chapter.

In mechanical lubricated contact systems where surface interactions do occur, these techniques offer the possibility to simulate different corrosion conditions under well-controlled electrochemical environments. This can be done for instance by the measurements of open-circuit potential (E_{oc} or E_{cor}) during a corrosion-wear test, measurements of cyclic potentiodynamic polarization curves under wear and pure corrosion conditions, records of the current-induced applied potential resulting in an infinitesimal disturbance of the surface (potentiostatic control), measurements of current-induced single step anodic potential pulse (for the study of film repair, or repassivation), measurements of electrochemical noise for on-line tribocorrosion monitoring, measurements of electrochemical impedance, measurement of the linear polarization resistance (LPR), etc.

Since most of the undertaken studies on chemical degradation of glassy materials involve the use of cyclic potentiodynamic polarization method, it would be interesting to outline the conditions and limitations of this electrochemical technique for its use in tribocorrosion.

3.2.1. Measurements of cyclic potentiodynamic polarization curves

The susceptibility of metals to localized corrosion is usually expressed by the breakdown potential, E_b, or designated to, as the pitting potential, E_{pit}, and the repassivation potential, E_r or termed the protection potential, E_p. At the breakdown potential, localized corrosion starts. The E_{pit} of a metal is often associated with the potential at which the current density suddenly increases and with the breakdown of its passive surface film. The higher the potential (more noble), the less likely the alloy is to cause the initiation of localized corrosion. At the repassivation potential, pitting stops.

A cyclic potentiodynamic polarization technique can be employed to determine both E_b and E_r of a corrosion system under wear (i.e., tribocorrosion) and pure corrosion conditions.

Many researchers [15–21] have successfully used the potentiodynamic anodic polarization technique to study the corrosion, the wear-corrosion synergism, and the tribocorrosion behavior of metallic alloy systems including the BMGs in various electrolytes [17–21].

ASTM G61-86 provides a procedure for conducting cyclic potentiodynamic polarization measurements [22]. By convention a cathodic current is negative whereas an anodic current is positive [23].

This technique uses a typical three-electrode system (WE working electrode or metal being investigated, RE (SHE, reference electrode), and CE (Platinum or Graphite, counter-electrode)) controlled by a potentiostat as shown in **Figure 2**. The potential, which applies to the WE, is usually swept from the active (cathodic) direction to the noble (anodic) one, while tracking the current density continuously until it reaches a selected value of current density, where after, the scan is inverted in the active direction, until the hysteresis loop closes or until the corrosion potential is reached. The potential of the WE can be considered as the "driving force" of the corrosion system, while, the anodic current density can be regarded as proportional to the corrosion rate of the WE.

A typical plot of polarization curve generated by this method as E–log (i) is shown in **Figure 3**. If a specified material is susceptible to localized corrosion, a hysteresis loop as shown in **Figure 3** will be observed as the potential scan is reversed. Otherwise, a uniform corrosion takes place in the transpassive or oxygen evolution region. Note that the larger the area of the hysteresis loop, the lower the ability of the metal to repassivate.

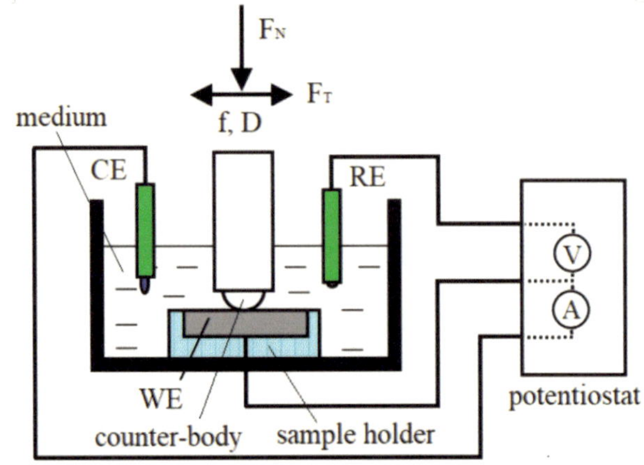

Figure 2. Schematic view of a tribocorrosion experimental set-up under potentiodynamic polarization conditions. Potential (*E*), and current density (*i*) measurements are performed on a working electrode (*WE*) sliding against a counterbody ball (unidirectional reciprocating sliding, sphere-on-flat) with respect to a *RE* reference electrode (e.g. Ag/AgCl (3 M KCl)) *via* a *V* voltmeter and *CE* counter-electrode (platinum or graphite) *via* an *A* ammeter respectively. F_N normal force, F_T tangential force, *f* sliding frequency, *D* displacement amplitude.

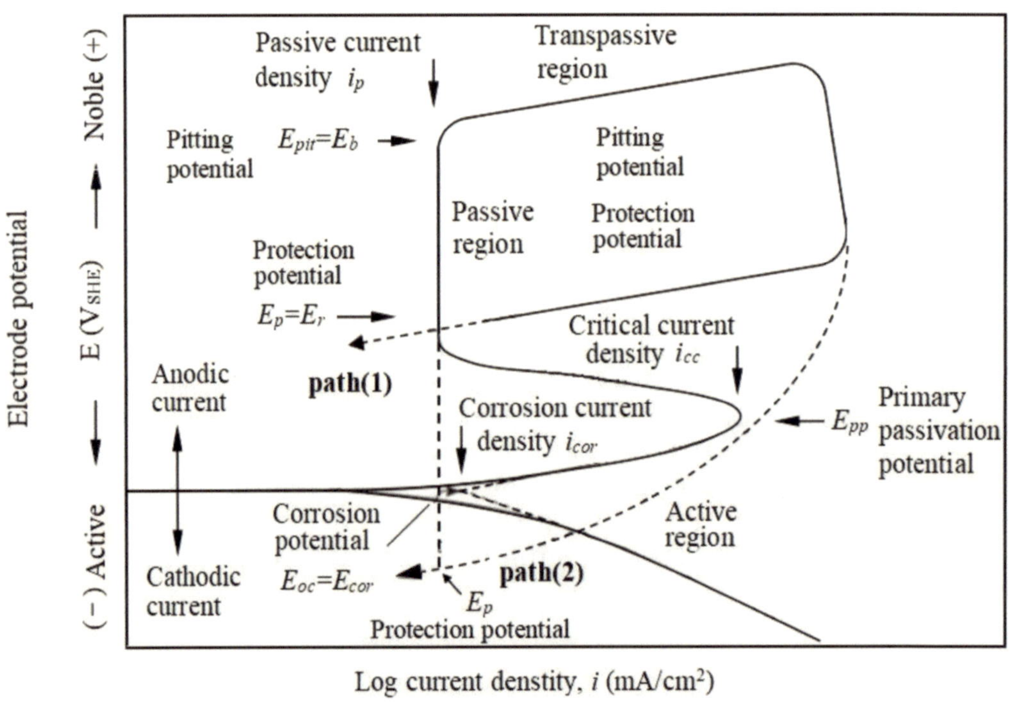

Figure 3. Schematic view of a hypothetical cyclic potentiodynamic polarization (cathodic and anodic) plot for determining localized corrosion parameters.

The approach to be considered for getting a quick picture of the electrochemical behavior of a metal under polarization is to identify some of the key corrosion parameters directly from the cyclic potentiodynamic scan mode (**Figure 3**), mainly:

- The corrosion behavior of a metal is assessed at E_{cor} (or E_{oc}) and other potentials.
- The process is designated as anodic polarization when the potential is above the E_{cor}, and cathodic polarization when the potential is below the E_{cor}.
- The corrosion potential, E_{cor}, corresponds to the potential at which the current density approaches or theoretically equal zero ($i \approx 0$).
- At cathodic potentials lower than E_{cor}, the cathodic reactions takes place (H_2 evolution or O_2 reduction prevails).
- The extrapolation of the experimental anodic and cathodic branches of the polarization curves, as shown in **Figure 3**, indicate the corrosion current density, i_{cor}.
- The primary passivation potential (E_{pp}) corresponds to a potential positive to which passive surface layers are formed, and at which a maximum corrosion current density is reached coinciding with the critical current density (i_{cc}, current requisite prior to the formation of surface layers).
- The region, where the current density remains approximately 10 mA.m^{-2} (i.e., passive current density, i_p of the electrode at E_r) as the potential increased, is called the passive region. This corresponds to the stabilization of a passive film. A material that exhibits a high resistance to localized corrosion in a specified environment form an adherent, and dense, nonporous, thin passive oxide/hydroxide film, which protects it from high corrosion rates.
- The protection potential E_r (or E_p) is the potential at which passive layers are stable and protective.
- At the potential interval, ($E_b - E_r$), pits are initiated and propagated. In fact, the material may suffer from localized corrosion when the potential is higher than E_r.
- High values of both pitting overpotential ($\eta_{pit} = E_b - E_{oc}$), and protection overpotential, ($\eta_p = E_r - E_{oc}$) may indicate a strong resistance of the metal to pitting at E_{oc}. High values of both (η_{pit} and η_p) are desirable to reflect high values of E_b and E_r relative to the E_{oc}.
- At potential above E_b, transpassive region starts. The local breakdown of the passive film continued driven by a rapid active dissolution (e.g., case of some stainless steels immersed in chloride solutions), and usually at more elevated potentials, oxygen evolution may occur (*cfr*. Pourbaix diagram, *E*-pH).
- In the reverse scan, as in case of path (1), the localized corrosion will not be triggered at E_{oc}; in this case, the working electrode will not suffer pitting under free E_{oc} conditions. If the path (2) is observed, where E_{pp} is below E_{oc}, the working electrode will undergo pitting corrosion on the surface defects or after E_{oc} incubation periods.

Many of the foregoing determined corrosion key parameters are based on empirical observations. As with any empirical method, it is perplex with many questions about the extent of its validity. The power of this technique should not be over-estimated, since the values of E_b and E_r are subject to change by a number of factors. Typically, for instance, environmental changes (e.g., temperature, pH, reagent as chloride ions) will influence these values drastically. Moreover, this method can be function of scan rate, pit size or depth, polarization curve shape, and specimen geometry, which can affect the accuracy of these electrochemical parameters (E_b and E_r) [24]. For example, if the scanning rate is too high, the E_b usually has a higher value than the correct value. This is due to the potential dependence of the induction period (time required for localized corrosion to initiate) which is long at low potentials and short at high potentials. To overcome this issue, the E_b can be measured by using a stationary potentiostatic method, in which the electrode is polarized at a constant potential and the time dependence of the current is measured. If the current starts to increase with time at a particular potential, then this potential is the correct E_b. Due to these uncertainties, this technique should be considered as qualitative and should never be used alone. In fact, it is often desirable to use at least two different types of corrosion monitoring devices whenever possible to weed out spurious or inaccurate readings.

A complete discussion regarding DC and AC electrochemical techniques for the measurement of corrosion rates is beyond the scope of this chapter. A more comprehensive treatment of this subject area may be found elsewhere [9, 11, 22, 65].

3.3. Aspects of tribocorrosion

Wear is an unavoidable and a potentially serious problem in all areas of engineering. In particular, wear due to tribocorrosion is reflected by a loss of material from the exposure to corrosion of contacting solid surfaces and in relative motion. Designers and engineers who have to make optimal decisions in situations where tribocorrosion considerations are significant, need to know "how long will a component last?" To solve this question, numerous models have been developed so far to distinguish this material loss due to tribocorrosion. These models usually correlate a wear volume or a wear rate with physical and geometrical quantities. Various expressions have since been attributed to this material loss, of which the material loss can be defined in terms of weight, volume, surface, depth, width or even charge density or current density, per unit hardness, per unit frictional dissipated energy (work due to the tangential force), per unit input energy (work due to the normal force), or even per unit sliding distance, or sliding time, sliding frequency, contact frequency, etc. It becomes readily understandable of the complexity of comparing results between the various wear data published so far. It is expected then that the terminology in this field is rather uncertain, and it will remain so for a certain time, hence the need for a specific standardization, despite some recent progress made in this area [25].

One of the earlier attempts to predict the wear rate or wear volume loss of a material in sliding contact is the commonly Archard wear criterion [26] used during the second half of the twentieth century. That criterion is usually expressed as follows,

$$W_v = k\frac{F_N}{H} \quad \text{or} \quad W_r = kA_r \qquad (1)$$

where, W_v and W_r represent the volumetric loss (assigned as total volume of wear debris produced), and the wear rate (usually expressed per unit sliding distance) respectively. k is the dimensionless Archard wear coefficient, A_r the real area of contact, F_N the applied normal load, and H the hardness of the worn material.

This equation was originally used for the case of adhesive wear [27, 28], then it was extended to more cases including that of tribocorrosion. This is because the k parameter in the Eq. (1) exclusively remained the only flexible parameter consistent with the case to which the wear may originate. For example, unidirectional sliding of mild steel against mild steel without any lubricant has a k of 10^{-2}, whereas, for stellite sliding against tool steel, k is 10^{-5} [29]. Even more confusing is that, according to literature, the Archard wear coefficient can vary by two orders of magnitude for the same couple of materials just due to a slight change in load or speed [30]. These findings should be taken with precautions in view of the number of empirical error cases reported with respect to the wear reproducibility and validation of test methods. Friction and wear properties are often considered as subjects of poor accuracy in comparison with materials intrinsic properties. Indeed, comparative round robin studies on the topic have shown that the reproducibility of wear derived from different inter-laboratories with the same material pairing was often very poor [31]. Using the wear track width, the scattering was roughly 50% whereas the scatter in the wear coefficient was over three orders of magnitude. No clear correlation was found between a single and constant parameter (type of tribometer, normal force, and sliding velocity) and the wear rates measured in inter-laboratories. Interestingly, a good convergence was found between the wear volume loss and the energy dissipated in the tribocontact zone [32–36]. This can readily be explained by the fact that the dissipation of frictional energy is one, among others, of the main causes of triboelectrochemistry, playing an essential contributing role in wear mechanism, in this case entailing an acceleration (e.g. chemical wear rates) or modification of tribochemical reactions. The yielded frictional heat between interacting surfaces leads to a stationary rise in temperature at surface contact asperities and flashes. Furthermore, such frictional energy can take the form of high quantum excitations with short lifetime of surface and bulk sites due to the mechanochemical forces involved during the sliding process. Those excitations are also responsible for the occurrence of triboluminescence and triboelectricity.

Tribologists nowadays are seeking for an agreement due to the fact that there is an unavoidably need to address more fundamental research towards the establishment of an original formulation or a universal methodology to define a "wear criterion" in order to better understand the complexity of the wear process in a tribocorrosion test. Although, this aim has not yet been achieved, a fair amount of progress has been made on this matter-oriented approach. This remains so far valid only as part of the case-by-case study.

To conclude, research must focus on establishing an approach that emphasizes the nature of the dependence of the mechanochemical wear rate (output) on the energetic aspect of sliding friction, the electrochemical aspect of the exposure of bare metal surface, and the transformation of the subsurface material (input). This usually should incorporate materials properties, and behavior. If it does, this could be very useful to solve most the issues and difficulties encountered in the specific field (e.g., cases involving the failure of mechanical systems related matter at any given time), and thereby leading to a better improvement of the reliability life of selected material and design technologies when adopted in mechanical articulations. Further, this could predict materials performance in an environment where tribocorrosion plays a significant role.

In tribocorrosion phenomena, where tribological contacts are exposed to corrosive environments, such as aqueous lubricants, the contact materials are subject to both mechanical, and chemical/electrochemical solicitations, which contribute to material removal from sliding surfaces. The rate of material degradation/removal cannot be predicted simply by adding the wear rate in absence of corrosion to the corrosion rate in absence of wear. The reason is that corrosion and wear do not proceed independently and synergistic effects usually (but not always) result in accelerated material degradation (tribocorrosion). In that respect, theoretical models have been developed so far with respect to mechanical, chemical, and electrochemical factors and their mutual interactions, and which can be tested under well-controlled experimental conditions. In general, modeling has followed either an empirical or a mechanistic approach. The empirical approach is based on the independent measurement of material loss due to wear and corrosion. These parameters are summed up and compared to the material loss due to tribocorrosion. The difference between the two is termed synergy (ΔW_{syn}). A general equation for this approach is of the form [37–39],

$$W_{tot} = W_{mec} + W_{cor} + \Delta W_{syn} \tag{2}$$

where, W_{mec} represents the material loss due to wear measured in the absence of corrosion, and W_{cor} is the material loss due to corrosion only without any influence of mechanical wear.

Although, the empirical approach is technically feasible which allows for the ranking and the performance of materials based on their resistance to tribocorrosion in engineering systems, it is still time-consuming, quite economically not justifiable in the long-term, and furthermore, it integrates a synergy term, which has no physical meaning.

The advantage of a mechanistic approach is that it leads for a better understanding of the physical processes involved in tribocorrosion by incorporating the notion of synergism into the mechanical and electrochemical terms. Many factors can be responsible for the mutual dependence of mechanical and chemical material removal in a tribocorrosion system. For example, local abrasion of the passive film can lead to wear accelerated corrosion due to rapid dissolution of the locally depassivated metal surface, followed by repassivation [40]. The abrasive action of hard oxide particles formed by corrosion can accelerate the mechanical metal removal by wear [41]. The plastic deformation of the surface layer of a rubbing metal can lead to a transfer of material to the opposite body resulting in a reduction of the corrosive wear rate [42].

Therefore, it is important to distinguish material loss due to chemical or electrochemical oxidation (i.e., wear accelerated corrosion) from material removed due to mechanical wear (i.e., mechanical material removal from the sliding contact). The former arises from the fact that an asperity sliding on a material surface produces a fresh wear track zone of clean bare material (i.e. metal), which is usually more susceptible to corrosion than the same surface subjected to free corrosion under no mechanical plastic contact or sliding conditions. The effect of repeated sliding may cause the removal of metal particles by asperities burrowing beneath the surface [12, 42, 43].

Therefore, the overall wear volume due to tribocorrosion, W_{tot}, can be defined as follow:

$$W_{tot} = W_{che(wac)} + W_{mec} \tag{3}$$

where, $W_{che(wac)}$ is the electrochemical contribution to wear; it is termed wear accelerated corrosion and it reflects the material loss due to corrosion in the presence of wear. W_{mec} is the mechanical wear, and it reveals the material loss due to wear in the presence of corrosion, and which can be related to processes as that for the formation-ejection of oxide debris, oxide layers or any corrosion products, and plastically detached metal.

W_{tot} can be determined by measuring the volume of the wear scar post-experiment using, for instance, a laser non-contact profilometry or by on-line measurement of the rate of moving down of the counter-body (e.g., a pin) on the surface wear track during sliding. The latter method has the advantage of recording an instantaneous wear rate, but it would only be applicable if no significant amount of solid reaction products (such as third body particles) accumulate in the contact zone during the tribocorrosion experiment. Under potentiostatic control, the electrochemical term ($W_{che(wac)}$) can barely be related to the anodic corrosion current ($I_{a,tribocor}$) measured under mechanical sliding wear (occasionally by subtracting the background current) using Faraday's law. The amount of anodically oxidized metal under such conditions is calculated as follows [12]:

$$W_{che(wac)} = \frac{M.q}{\rho.z.F} \tag{4}$$

where, $W_{che(wac)}$ is the volume of the metal transformed by anodic oxidation in a triboelectrochemical test. $q(t) = \int I_{a,tribocor}.dt$, is the electric charge generated during that transformation process, which is obtained by integrating the measured current $I_{a,tribocor}$ over the time of the triboelectrochemical experiment, M is the atomic mass of the metal, z is the valence for oxidation reaction, F is the Faraday constant (96,480 C/mol) and ρ is the density of the metal.

This equation is credible and independent of whether the anodic oxidation leads to the formation of dissolved metal ions or solid reaction products, such as oxide films.

It is worthwhile to note that few assumptions must be met in order for the Eq. (4) to be used [12, 40], namely:

- The measured current must be equal to the anodic partial current for metal oxidation, which means that cathodic partial currents due to the reaction of oxidizing agents must be negligible. This can be performed by anodic polarization into the passive potential region.
- The charge number z for the oxidation reaction must be known [12, 40, 42].

The mechanical wear (W_{mec}) is taken as the difference between the total wear volume W_{tot} and the chemical wear volume $W_{che(wac)}$ determined from the electric charge.

3.4. The causality of wear-corrosion synergism

The synergy between wear and corrosion has recently attracted increasing attention to improve materials used in systems where tribocorrosion plays a role. However, wear-corrosion synergy still seems to be a developing topic of discussion as no convincing expression is available yet.

A positive or a negative (antagonism) synergistic effect can occur in most cases where surface interactions interfere with tribocorrosion phenomena. It intervenes especially (in a positive way) by increasing the wear (volume) when the mechanical process affects the electrochemical process and *vice versa*. In these situations, the total wear (volume loss) will be very different and greater than the sum of the mechanical wear in the absence of corrosive environment and the loss of material by corrosion in the absence of any mechanical stress. A negative effect of synergy, however, will occur when the total wear is less than the sum of the two protagonists taken individually and independently, namely wear and corrosion.

Madsen et al. [37, 38] critically reviewed the measurement of wear-corrosion synergism, and proposed a group of penetration rate equations to quantify the wear and corrosion processes and the wear-corrosion synergism. Their results showed that the wear-corrosion synergism is of great extent for alloys, such as AISI 316 stainless steels, which depends on the formation of a film of passive layers for their corrosion resistance, sometimes only a few atom layers thick, resulting from an interaction between the material and the surrounding environment [37, 38]. On the contrary, the synergism was limited for alloys, such as low alloy steels (e.g. amorphous steels), which do not depend on the presence of a passive film for their resistance to a corrosive environment. The causality of this synergistic effect has been explained in part for some passive materials. The presence on their surface of passivation layers and the ability of their surface, even in a deformed or partially damaged state, due to the sliding contact by the counter-body or a third-body, to be rehabilitated (by forming reaction layers with a thickness of a few nm, such as oxides, solid precipitates, adsorbed layers or passive surface films) is at the origin of the increment of that synergy. Dense oxide layers, precipitates, or passive films may play a protective role in isolating the underlying metal from a direct contact with the surrounding corrosive environment, thereby protecting the metal from a corrosion increment, but not necessarily their mechanical wear. In particular, one of the possible explanations could be related to the mechanical or chemical shear strength of these formed layers. This can likely be one of the causes of the incremental or decline effect of the wear-corrosion synergism of passivating metals. This is mainly true in the case of stainless steels and other alloys containing chromium. Their passive surface film formed in the ambient air or in contact with an aqueous solution has a thickness of a few nanometers but gives them a high resistance to corrosion. The sliding of a hard counterbody material on such a surface is likely to damage that passive film what is known as a "depassivation" process by which the bare material is exposed to the corrosive environment [8, 10, 13, 32, 44]. Various but essentially electrochemical processes can then compete on these bare surfaces [8, 10, 13, 32, 44], namely:

- The dissolution of the metal in the corrosive medium;
- The formation of a new compound that may contribute to the breakdown process;
- And the restoration of the protective film known as "repassivation" process.

http://dx.doi.org/10.5772/intechopen.78233

Most amorphous alloys, such as bulk metallic glasses (BMGs), usually do not depend on the presence of a protective surface film to exert a corrosion resistance. Therefore they can be expected to claim showing a negative synergistic effect, hence their opportunity to be selected as potential candidates over other passive alloys (such as stainless steels) in systems where tribocorrosion plays a role.

Celis et al. [10, 13] identified five mechanisms in tribocorrosion, which could explain the synergism noticed between mechanical and chemical factors acting on contacting materials, mainly:

1. The debris can speed up or reduce wear compared to what happens in the same environment where debris does not exist like e.g., in sliding contacts polarized at a large cathodic potential;
2. A galvanic coupling is established between the worn (active) and unworn areas (passive). It accelerates the anodic dissolution in the area where the metal is depassivated;
3. A galvanic coupling may be established between the two contacting counterparts;
4. An accumulation of dissolved species may take place in the liquid surrounding the contact. This may render the medium chemically or electrochemically more aggressive;
5. The mechanical loading in the contact area and its nearby zone may cause a work hardening of the materials. This work hardening can alter the kinetics of corrosion and/or repassivation processes.

3.5. Passivity breakdown

Most of engineering and biomedical metals and alloys oxidize and frequently passivate "spontaneously" in contact with the ambient air, and with suitable aqueous media to form "natural" thin passive surface oxide films. Alternatively, alloying is also considered to be, inter alia, the most extensively used method for enhancing the passivity of base metals [45]. Passive films that growth on most surface alloys are found to be of two types: discontinuous, and continuous [46]. Discontinuous films are porous and are formed at the metal/solution interface from the reaction of metal cations with species in solution. They have a thickness of up to 1 mm and are less protective. Continuous films, on the other hand, are tenacious and thin (nm's to μm's thickness range). They have high resistances ($\geq 10^6$ ohm.cm^{-2}) and support a high electrical field. They are called barrier protective layers, which serve to prevent current flow, and corrosion (i.e., dissolution). Commonly, the passive film on most metals and alloys consist both continuous, and discontinuous layers, with discontinuous film forming the outer layer, and the continuous film forming the inner barrier protective layer. A typical example of a bi-layer passive film formed on Fe-based amorphous alloy and interacting with anions adsorption in various aqueous solutions [47] is depicted in **Figure 4**.

The approach that links the improved resistance of amorphous alloys to their ability to promote amorphous passive oxide formation is well accepted by the scientific community. In high-temperature gas working conditions, vitreous or amorphous oxides offer improved

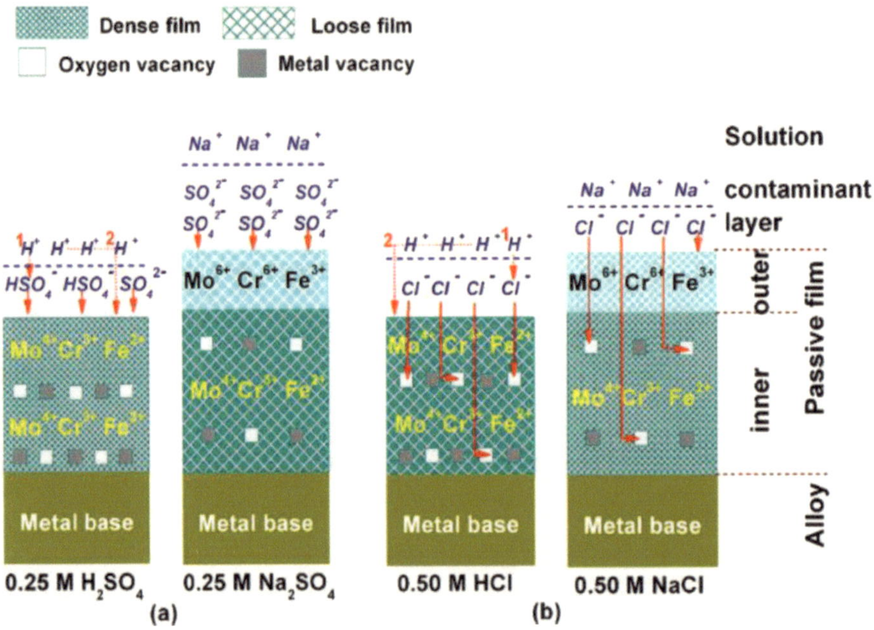

Figure 4. Schematic diagrams of anions adsorption and characteristics of passive films (bi-layer structure) formed on Fe-based amorphous alloy in (a) sulfate and (b) chloride solutions. (reproduced from Wang et al. [47] with permission from Elsevier Science).

oxidation resistance due to the absence of oxide grain boundaries, which provide a rapid diffusion path for concentration gradient-driven ion movement [46]. In aqueous solutions, ion transport is mostly driven by the electric field across the oxide film. The lack of oxide grain boundaries may lower ion migration rates, rendering the passive film more protective [46, 48].

It is worthwhile to mention that if the passive films on metals, like iron, nickel, and chromium, remained intact, then the corrosion current flowing across the interface under most industrial conditions would be of the order of 0.01–1.0 $\mu A.cm^{-2}$, corresponding to corrosion rates of approximately 0.15–15 μm per year [49]. For most practical situations, metal loss rates of this order are of no concern, so that our automobiles, bridges, aeroplanes, and industrial systems would last for periods extending well beyond the current design lifetime. Unfortunately, passive films do not remain intact, and corrosion rates of many orders of magnitude greater than those indicated above for fully passive substrates are commonly observed, particularly if the attack occurs locally.

Passivity breakdown can occur for a variety of reasons, including mechanical straining of the substrate metallic alloy, the frictional dissipated energy in tribological contacts required for micro-cracking, the presence of thermal stresses within the oxide due to differences in thermal expansivity, compressive stresses in the oxide growth (Pilling–Bedworth ratio), fluid flow and cavitation, transpassivity polarization, and chemically-induced phenomena. In particular, the rapid transport and accumulation of cation vacancies at the oxide/metal glass interface is one theory accounting for oxide breakdown [50]. Moreover, vitreous oxides on amorphous alloys perform well due to the enhanced bond flexibility, because the vitreous or amorphous material can rearrange to accommodate lattice mismatch and strain between the oxide and the metal [48, 51]. As a result of this flexibility, almost all surface atoms can bond with oxygen or OH without requiring an optimal epitaxial relationship between the ordered metal substrate and oxide. Intolerable changes in oxide/metal misfit strain with halide incorporation is another theory accounting for the rupture of protective oxide films on metals [52].

In the case of crystalline alloys, such as the highly passivated stainless steels, it should be noted that the improved corrosion resistance associated with the addition of 18% Cr to crystalline iron is attributed, in part, to a change in the protective oxide structure from a well-oriented spinel structure at 0–12% Cr to a non-crystalline structure at 18% Cr [48]. This disordering has been demonstrated by low-energy electron diffraction [53] and scanning tunneling microscopy [54]. A retardation in ionic transport may occur, because non-crystalline films have fewer defects or grain boundaries to enhance ionic movement. In summary, desirable amorphous oxide/amorphous alloy properties include defect minimization, film ductility, bond flexibility, and efficient, rapid film repassivation, which all contribute to improved corrosion resistance [48, 51].

4. Consensus on the need of materials for triboelectrochemistry systems

The development of tribocorrosion resistance requirements has resulted in the development of a much larger number of materials that did not exist before, especially metallic alloys. In turn, such a development of materials has multiplied case studies, and increased the number and the diversity of corrosion-wear problems. Accordingly, the resolution and the nature of corrosion-wear problems are intimately related to the choice of materials. This does not mean that the tribocorrosion resistance is necessarily the determining parameter in the choice of a given material. Such a choice must, in fact, make it possible to fulfill at best one or more technological functions, and it is quite obvious that, under such conditions, the mechanical properties of the materials, their properties of implementation, their price or their availability are in many cases, parameters are just as decisive in the choice as their only resistance to tribocorrosion.

5. Evolution of the solutions to triboelectrochemical problems

The choice of metals and their alloys seems to be more relevant for designing materials over plastics, foams, polymers, and natural materials if the intended application requires a limiting risk factor of deformation and flexibility, hence interest on their application to load-bearing structures. Alternatively, the development of novel metallic-glass-matrix composite materials has open up the opportunity for alloy design and innovations in new bulk metallic glass (BMG) materials to withstand deformation and flexibility that cannot be achieved by traditional metals, or casual BMGs making them attractive for various tribological systems (e.g., journal bearings), and mechanical engineering applications.

The ultimate resolution combining selection and material design is a consensus between economical parameters and technical qualification. Generally, this requires three essential steps, namely:

1. Identification of the requirements (e.g., design considerations);
2. The choice of materials assigned for the function of interest, including, inter alia, mechanical and chemical risk factors;
3. The election of the least expensive material.

This third step in the choice of a material emphasizes the problem of economic compromise. It is thus brought to compare investment costs (resistant but expensive material) and operating costs (costs of failures, replacements, and stops they may cause). The relative weight of these two types of cost has slowly changed. The current trend is often to prefer high but predictable investment costs and to minimize operating costs when these are too difficult to predict. In addition, in a competitive economy, short-term cash optimization is increasingly being replaced by optimizing long-term profitability. In turn, this may well favor investment, especially in metallic glass alloys whose resistance to mechanical and chemical constraints (i.e., to tribocorrosion) is optimal.

In what follows, we will particularly focus on the second step above, namely the behavioral of BMG materials with regard to their wear and corrosion testing, and in particular to their mutual coupling effect (tribocorrosion).

6. Towards the needs in tribological and electrochemical testing to meet the requirements of effective usage of metallic glass materials

Actually, despite the cost for developing new BMGs of desirable types and compositions, the examination of their appealing properties remains a persistent issue. Although, wear and corrosion are an important topic, they have never received the attention they deserve. Many tribological and electrochemical aspects of BMGs have not been characterized yet or are not well assumed, and the actual deformation and failure mechanisms are not fully understood. Albeit, localized corrosion, fatigue, wear, and fracture all have been reported in almost every BMG corrosion or mechanical study published to date (see next chapter). Though, theoretically speaking, this fact is unexpected for an ideally homogeneous material. Therefore, an overview of the material-oriented approach to resist the mutual and opposite interactions between the main actors of tribocorrosion involving both mechanical (friction, wear), and electrochemical (corrosion) phenomena is given below.

7. Material properties, which influence surface interactions in a triboelectrochemical contact

Fundamentally, tribocorrosion depends on the dominating deterioration mechanism of interacting surfaces in chemical environment and under relative motion conditions [10, 55]: *viz.* wear, corrosion, and their mutual interaction (synergism).

Modern research has established a consensus on four main forms of wear, namely, chemical wear (i.e., corrosion and corrosive wear), adhesive wear, abrasive wear or surface fatigue wear [56]. Each process of wear obeys its own laws and, to confuse things, repeatedly one of the modes of wear acts in such a way as to affect the others, hence of the complexity of wear. Typically, there is a combination of wear mechanisms in a mechanochemical dynamic contact. In that respect, the classification of wear mechanisms remains a matter of debate among the scientific community of researchers and authors. Albeit, the terminology used by Burwell in 1957 [56] to describe wear is simple and rational, that of seeking out the primary cause of each form of wear. To avoid any further issue regarding the nomenclature in this field study, all wear mechanisms should be referenced to the ASM standards [57, 58].

Other forms of wear can be found in the literature depending upon the contact configuration (e.g. unidirectional and reciprocal sliding and/or rolling, rolling with slip, etc.). Wear in these contact geometries is reported to as erosion wear, fretting wear, sliding or rolling wear, impact or slurry wear, etc. This is one of the approaches that judges wear by the consequences of the conditions of a tribological contact *vis-a-vis* its environment (e.g., reactivity of tribosurfaces with the environment, the contact system configuration, etc.). Such wear descriptions are all-technical and do not represent wear mechanisms in a scientific manner.

According to the recent critical review on the quantification of wear made by Meng and Ludema [55], several models have been proposed to explain various phenomena of wear. The authors have clearly enumerated 182 equations with 625 variables for explaining wear processes. This clearly shows that wear is not a material property, but rather a material system response. Wear can change drastically even as a result of a relatively small change in dynamical, environmental or material parameters forming the tribosystem. Indeed, wear rates change promptly (10^{-15} up to 10^{-1} $mm^3.N^{-1}.m^{-1}$) depending upon the conditions in which the materials are exposed to (tribological system, corrosive medium, loading contact parameters, etc.) and the choice of these materials [26, 59–61]. The combination of these two main factors, namely the operating conditions and the choice of materials, are the primary keys for monitoring the wear of materials (modes and rates) exposed to normal working conditions. Optimal solutions have been recommended as a means of meeting these requirements, are the wear maps that predict both modes and rates of wear of materials [8, 62]. A wear map or chart can be considered as one of the best descriptions of tribological/tribocorrosion conditions and as useful strategy in the design of mechanical systems (tribosystems) and for the selection of materials to be used in a wide range of operating conditions.

The wear volume loss measured during or after the end of an operating tribological test provides useful information in characterizing wear. Generally, there are three typical types of

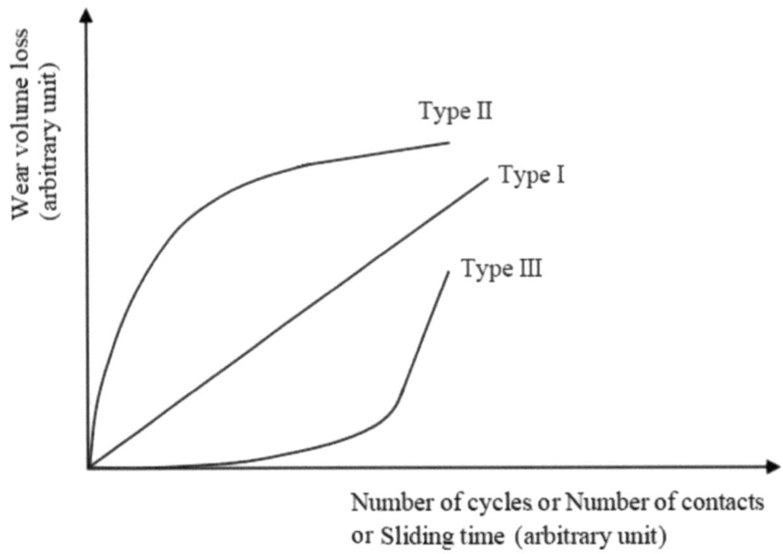

Figure 5. Three typical wear curves in repeated sliding contacts.

wear volume curves as shown in **Figure 5**. Type I shows a constant rate of wear throughout the process (likely one ideally mechanism monitoring wear). Type II shows the transition from an initially high wear rate to a steady-state low wear rate. This type of wear behavior is often observed in metals and metallic alloys [63]. Type III shows a fatal transition taking place from an initial low wear rate up to a high wear rate (e.g. fatigue wear by fracture mode). This is frequently encountered when using ceramic materials [64]. The total number of sliding contacts, before any catastrophic wear can occur, reflects the period for which the crack initiation takes place, and this latter depends on a number of material and system parameters, mainly surface roughness, material properties, and sliding conditions.

When considering the tribocontact system as a whole, including the entire design configuration and components, and specific for a given application either in laboratory or in the field, it is not possible to increase the strength of the solid material *vis-a-vis* one of the wear mechanisms in general. Nevertheless, the selection of suitable materials should always be established based on a more knowledge and a better understanding of the conditions in which the solid material is subjected to (e.g., surface interactions in aggressive medium, and loadings).

In the following sections, an overview of the desired material properties and trends towards how wear and corrosion resistance can be achieved is given.

7.1. Materials oriented approach to act in opposition to corrosion and corrosive wear

7.1.1. Corrosion

Corrosion may be defined as "…an irreversible reaction of a material with the environment, which usually (but not always) results in a degradation of the material or its properties…"

http://dx.doi.org/10.5772/intechopen.78233

Therefore, there are several aspects of corrosion: the material, the environment, and the material properties. Considerable information is available in textbooks [9, 45, 65]. This general definition of corrosion includes the physicochemical (oxidation/reduction) reaction processes taking place at the surface of widely varying material classes (metallic, ceramic and organic), such as localized corrosion cells in some Fe-based metallic glass ribbons used in waste water treatment, the degradation of dental amalgams or metallic restorations by galvanic corrosion, polymers by ultraviolet radiation, and the chemical attack of refractory bricks during steel-making.

The severity of a corrosive medium depends on a variety of parameters. Nevertheless, the following four main parameters can generally be selected, namely the pH of the medium, the presence of chlorides (and other alkyl halides), the oxidizing power, and the temperature. The corrosion resistance properties can then be characterized directly by the limits of use of the materials, which can be expressed, for example, in terms of maximum temperature in-service or maximum concentration of use. In real service and under normal conditions, the understanding and control of corrosion are based on the electrochemical interpretation of corrosion phenomena and the consideration of the relative ranking scales of materials in order to select, by successive approaches, the materials best suited to each application of interest.

Since these materials may interact in a complex way with the environment that they are used for, the assessment of the reactivity of their surfaces with respect to their environment, and the evolution of that reactivity with time greatly accounts for their corrosion performance. Usually, certain metallic materials gain their resistance from the formation of a thin, yet dense and adhesive oxide layer, which protects the underlying substrate from further oxidation (a passive layer). Thus, the potential at which this layer is growth and the rate at which it is built (passivity) and rebuilt after being mechanically or chemically damaged (repassivation) are decisive for its corrosion resistance.

On a theoretical level, passivity concerns, a priori, most metals (and alloys). However, only a small number of those metals actually allows a systematic usage of this property. Conversely, some metals resist to corrosion without any appealing to passivity. Some of them are chemically inert, or not very reactive; this is the case for example for noble metals, such as gold, platinum and iridium, which are not prone to oxidation due to their high standard electrode potential, and are hardly used in technical applications because of their excessive price and often poor mechanical properties, or even copper and nickel in non-oxidizing acids. Other metals are roughly covered with a protective layer of corrosion products; this is the case of steels exposed to atmospheric weather, copper alloys in natural waters, lead in sulfuric acid, or even, occasionally, the most ordinary steel in the presence of acidic gas medium, CO_2, or H_2S.

In total, metals and engineering alloys, including metallic glasses, actually used for their passivity are limited to Fe, Co, Nb, and Ta, stainless steels, alloys of Al, of Ni, of Ti, and of Zr, etc. It should be noted that some metals are less passivable than others are; especially this is the case of Fe, and Co, which do not reveal any sufficient passivation or oxidation. Not all oxide layers that form on metals are protective. If the oxide does not form a continuous layer on the surface of the metal, it will not be able to reduce the amount of oxygen reaching the metal surface, and thereby increasing the brittleness of the layer leading to further corrosion (*cfr.* Section 3.5).

7.1.2. Corrosive wear

A more obvious mechanism of triboelectrochemistry is the periodic exposure of fresh bare surfaces when sliding friction between surfaces occurs in corrosive liquids or gases. This results in reaction products mainly driven by chemical and electrochemical interactions. The surfaces of the materials are quickly covered by a scale of the reaction product, the oxide in the case of metals and metallic alloys, acting as a protective barrier layer. The thinner the scale, the faster the reaction, and the weaker the protectiveness.

In the case where these reaction products strongly adhere to the surface and behave as the bulk material, the wear mechanism should be almost the same as that of the bulk material. Otherwise, as observed in many cases in practice, the reaction products behave rather differently compared to that of the bulk material. The resulting wear is therefore very different from that of the bulk material, and it is thus controlled by the scales of reaction products (i.e. tribochemical reactions). In corrosive media, the tribochemical reaction at the contact interface is accelerated by the friction processes (*viz.* elastoplastic deformation, heating, micro-fracture, and successive removal of these scales of reaction products). In the case of metals, friction can cause extensive plastic deformation of a subsurface layer in the material; whereas in the case of ceramics, microfracture predominantly occurs. These strains lead to structural defects (microcracks, grain boundaries, vacancies, dislocations, etc.), thereby accelerating the diffusion of reagents through the protective scale. This results in the acceleration of the chemical reactions, which leads to a material removal from the contact interface. The resultant wear is called "corrosive wear". In air, oxygen prevails as a corrosive medium, and tribochemical wear of metals in air is usually called "oxidational wear". More detail about this type of wear can be found elsewhere [29]. The material removal rate in corrosive wear is governed by the balance between the relative growth rate and the removal rate, which determines the wear rate of the reaction layers.

It is worthwhile to note that in the absence of chemical reactions, the sliding surfaces experience mechanical wear, while in the case of the absence of sliding friction, the surfaces experience corrosion degradation.

7.2. Materials oriented approach to act in opposition to adhesive wear

Adhesive wear is the most common form of wear that exists when one solid surface material is slid over the surface of another (e.g. tribological/tribocorrosion contacts) or is pressed against it (e.g. loaded surfaces under bending conditions, fretting mode II, etc.). The removal of material takes the form of small particles which are usually transferred to the other surface, but which may come off in loose form [27].

In loaded or pressed tools, the tendency of contacting surfaces to adhere arises from the attraction forces, which exist, between the surface atoms of the two materials in intimate contact. In a presumed elastic contact, elastoplastic deformation of the asperities in contact takes place, which often results in a growth of the real contact area as the load increases [32]. This, in turn, leads to increasing the area for which the distribution of bonding and attraction forces may occur by forming an interfacial junction area along the contact, thus resulting in a strong adherence and promoting surface welding at a solid-state (*viz.* adhesion). The removal of the load breaks off most of the junctions, as a result of elastic spring-back. At this point, the chemical, plastic deformation and wear become clear in light of the formation of a series of grain-sized microwelds, microcracks, and material transfer.

The presence of adsorbed films containing water and other molecules derived from the air serves measurably to reduce the surface interaction of contacting materials. The effect of grease films, if present, however, is even more marked, and reduces, often by one or more orders of magnitude, the severity of surface interaction, and thus friction too. Clean and smooth contacting metallic surfaces are more prone to promote higher attraction than contaminated surfaces (e.g., oxides, corrosion products) and thereby increasing both the adhesion and the static frictional strength of interfaces (i.e. a certain minimum tangential force is required to produce motion).

Most non-metals essentially have the same chemical composition at the surfaces as they do within the interior. A handful of metals and alloys will form surface oxide films in air, however, and in other environments they are likely to do, for other films (e.g., nitrides, sulfides, and chlorides).

Under sliding conditions, the first stage involves the removal of the thin surface film (e.g. oxide layer) covering the surface of the material through a mechanical wear process. As the oxide layer is degraded after the first few sliding cycles, the distance between the two surfaces becomes in the order of the interatomic distance of the metal lattice, and metallic bonds are thus established (*viz.* direct adhesive bonding contact) between the two interacting surfaces (e.g., counter-body and underlying bare surfaces), thereby leading the contact surfaces to starting adherence by forming "microwelds" or "cold welds". When the adhesive bonding strength resists the relative sliding motion, large plastic deformation caused by dislocation is produced in the contact zone under the effect of compression and tangential shearing. The generation of slips along slip planes in the contact zone entails the formation of flake-like shear tongues and/or wedge-like shape, and are followed by a crack initiation and propagation in the combined fracture mode of tensile and shear in the contact zone area. Currently, the real actual contact area is made up of all areas of asperities welded to the surfaces and upon tangential sliding/shearing, the crack reached in the contact interface causes a mass separation of the asperities in the underlying softer material rather than in the interface, and therefore a wear particle from the bulk material is formed and eventually transferred to other surface.

The adhesive wear process is responsible for the initial accumulation of wear debris between the contact surfaces. Additional sliding cycles bring near-surface plastic deformation, new bonds often prevail and fracture occurs at some distance to the surface, additional wear and the potential formation of new oxide [7, 8, 10–13, 16, 32].

Under tribocorrosion conditions, the surface roughness, as well as the oxide film growth or chemical corrosion layers adsorbed on top of the surfaces, normally hinder the direct metal contact. However, the plastic deformation of the surface can result in the flattening of the roughness as well as the cleavage of the oxide layers. Preventing plastic deformation or solid-state attraction is therefore a feasible concept to avoid adhesive wear.

The role of the environment on the adhesive wear is of a great importance in tribological contacts. Under dry friction conditions, for example, often part of the energy is dissipated in the form of thermal heat, which boosts the surface reactivity and results in an increase of the temperature of the interacting surfaces. This, in turn, affects both the oxidation and the friction of the two surfaces in contact and entails material transformations (e.g. phase transformation, depletion zone, hardening, welding process, material transfer, etc.). Little part of the frictional work is done to overcome surface roughness, but most of the work being done in inducing shear displacement of the junction interface as well as the removal of bulk underlying previously welded material.

One of the interesting approaches used to explain the friction behavior of metallic oxides just below and above their melting temperature is the polarizability approach, initially introduced by Vesselin et al. [66], and later extended by Prakash et al. [67]. Thus, based on that approach, the lubricity of a wide variety of solid oxides at high temperature could be explained.

A correlation may exist between the dissipated energy due friction and the wear rate. Often a low friction induces a low dissipated energy, which may result in a low wear rate. However, this is not a universal law, and there are numerous examples showing high wear rate regardless of low friction.

Despite its omnipresence in almost all modes of degradation by sliding wear, adhesive wear is not necessarily preventable form of wear, nor is it the most dangerous since adhesive wear rates are usually fairly low. The mechanism behind this type of failure wear process is quite expected.

In general no equation can fully describe the adhesive wear process but the most widely used is the Archard equation (Eq. (1), see Section 3.3).

Adhesive wear can be addressed effectively by deliberate and intelligent choice of materials. Solid materials with high hardness or having undergone strain or work hardening (e.g. metals) lead to a reduction of plasticity of the surface. Alternatively, metals or alloys with a hexagonal or body centered cubic (bcc) crystal lattice are thus preferred to those with a face centered cubic (fcc) structure, since these show a high number of slip systems making them very ductile. Solid materials with covalent bonds are less prone to adhesion than those with metallic bonding. Adhesion, thus, can be effectively reduced by ceramic layers that can be produced by coating technologies or surface treatments. Nitriding of steels, for example, can lead to the formation of a nitride layer that reduces adhesion. If these layers, however, are damaged or removed by superimposed abrasive or cyclic loading, the underlying material become unprotected, and may favor adhesion. This often happens during tribological and tribocorrosion processes.

In case of metallic glasses, surface treatment or coating with material composites can be very useful to improving their hardness, and thus reducing their plastic deformation. A fine dispersion of hard phases into metal glass matrix for instance can effectively reduce adhesion.

After mechanical processing, they slightly protrude from the surrounding matrix and impede metallic contact.

7.3. Materials oriented approach to act in opposition to abrasion wear

Abrasive wear is defined as wear by displacement of material caused by hard particles or hard protuberances, which results in a significant plastic deformation of the softer surface material. Actually, this is the form, which occurs when a rough hard surface, or a soft surface containing hard particles, slides on a softer surface, and ploughs a series of grooves in it. The material from the grooves is displaced in the form of wear particles (debris) generally loose ones. When these wear particles are attached to one of the surfaces in contact, the phenomenon is called as two-body abrasion. Otherwise, it is known as three-body abrasion. The severity of abrasive wear depends on size and angularity of abrasive wear particles and the ratio between hardness of metal and the abrasive particles too [68, 69]. To be effective, either the hard particles or the harder surface must be 1.3 times harder than the softer surface material undergoing abrasion, which Hutchings [68], and Ludema [69] note is the difference of one unit on Mohs scale of mineral hardness.

In a tribocorrosion process, abrasion usually occurs when interlocking of inclined or curved asperity contacts at the contact interface takes place or when harder particles are introduced into the tribosystem between the interacting solid surfaces at the contact interface (e.g., through erosion (abrasive particles) or fretting (trapped in contact), etc.). These sharp and hard particles or asperities are then pressed onto the softer surface (usually being the investigated material), causing a plastic flow of the surface. During sliding-corrosion, the softer surface undergoes ploughing, entailing the formation of scratches, and abrasive grooves, which leads to a significant material removal (e.g. as volume).

It is likely that diverse modes of action contribute to the mechanisms by which abrasive wear occurs and proposed models include micro-cutting, micro-chipping, and micro-fatigue (e.g., due to cyclic loading-sliding). Other models have also been emphasized. Hutchings [68] quotes three common models for the occurrence of abrasive wear *via* plastic deformation, which can be categorized as cutting, ploughing and wedge-forming.

Note that the contact geometry in a tribological/tribocorrosion system remains, inter alia, a parameter either favoring or discouraging the manifestation of abrasion phenomena when a hard surface slides on another softer one or remained in intimate contact with it. In fretting or unidirectional reciprocating sliding conditions, for example, the use of a sphere-on-flat configuration can promote both trapping and ejection of the abrasive particles brought into contact between the two slid protagonist surfaces, whereas, in a flat-on-flat contact geometry, such abrasive particles may remain regularly trapped in the contact zone, driving various possible mechanisms as plastic flow, welding, scratches, grooves, etc. It is therefore recommended that the project designer or engineer consider the incidence that the choice of the geometrical contact system may have on the wear of the contacting materials.

The abrasive wear rate is defined in the same way as for the adhesive wear. Indeed, the Archard equation (Eq. (1), see Section 3.3), which was originally formulated to model adhesive wear, is commonly used for abrasive wear, although this is derived from a completely different set of material removal mechanisms. However, the validity of the Archard wear criterion is still questioned by the scientific community (see Section 3.3).

According to the Archard equation, a timely way to avoid abrasion is to increase the surface hardness of the component [70]. However, it should be pointed out that for a number of metallic glass composites and bulk glassy alloys, the wear rate may deviate and even do not follow the Archard equation [26]. Only a good combination of the hardness and the toughness taken together can allow the metallic glass to be wear resistant. A convenient way of bringing out this choice is by a series of figures or charts, in which one parameter of interest is plotted against another. The Ashby chart [71] plotted in **Figure 6** compares the normalized wear rate and the hardness for most of the common engineering materials including metals, technical ceramics, and polymers. In that figure, the wear-rate constant, k_a (MPa^{-1}) is defined as the ratio of the volume of material removed (m^3) to the distance slid (m) multiplied by the normal load (N). That quantity represents a measure of the propensity of a sliding couple for wear: if k_a is high this would correspond to a rapid or severe wear at a given bearing pressure.

The wear rate of metals are markedly hardness dependent, however, technical ceramics show nearly the lowest wear rate and the largest hardness over metals, polymers, and elastomers. Note how certain engineering materials lie roughly on a diagonal (dotted lines). Interestingly, the wear rate is strongly correlated to the hardness.

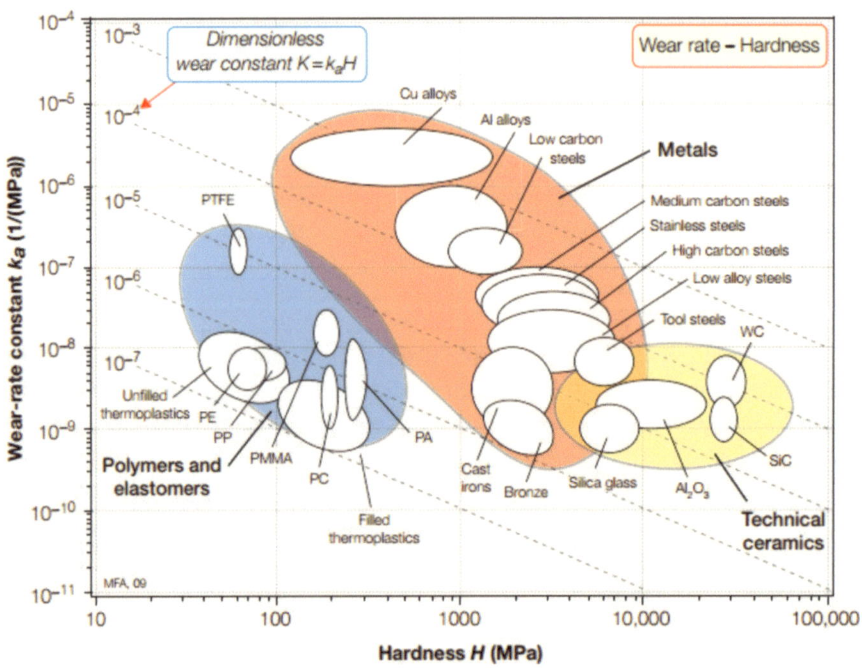

Figure 6. Ashby plot comparing the normalized wear rate k_a to the hardness H, here expressed in MPa rather than Vickers (H in MPa = 10 H_v). The chart gives an overview of the way in which common engineering materials behave. (reproduced from Ashby [71] with permission from Elsevier).

The hardness of a metal is more likely structure dependent, so that the hardness of an annealed metal may be lower than that of a work-hardened metal. The state of a metal in fully work-hardened condition corresponds to the state, which a metal surface might achieve after repeated sliding over it had occurred. Steels, for example, when hardened by quenching may be less prone to abrasive wear than soft-annealed. The application of this concept is, however, limited especially against mineral abrasives: metallic materials cannot reach the required hardness. Ceramics, on the other hand, are sufficiently hard but owing to their lack of toughness (**Figure 1**, see Section 2) and ductility they cannot claim to be eligible candidates as basic materials in many specialized fields. In addition, they are prone to micro-cracking during abrasive loading, which reduces their wear resistance. To counteract this issue, metallic materials are reinforced with hard phases. The tough/hard metallic matrix provides high ductility and strength, while the hard phases prevent indentation and grooving of the surface [72]. They are most effective if they are harder than the abrasive, and larger than the groove width. The hardness of precipitated-hard phases is raised by hard phase forming alloying elements. Their size usually decreases with the temperature of precipitation, i.e., primary, eutectic and secondary carbides. Tool steels and white cast irons belong to this group. Another group comprises metal matrix composites, e.g., Al-, Fe- or Co-based with a mixed hard phases, such as carbides (fused tungsten carbide, TiC), oxides (Al_2O_3, ZrO_2), cubic boron nitride or diamond. The matrix hardness needs to be sufficient to back up the hard phases and to minimize the indentation depth of the abrasive.

7.4. Materials oriented approach to act in opposition to fatigue wear

Fatigue (delamination) wear is defined as, "the removal of particles detached by fatigue arising from cyclic stress variations". The delamination or fatigue theory of wear was proposed by Suh [73], as an attempt to explain weaknesses in the Archard theory of adhesive wear [26].

Repeated cycles of contact are not necessary in adhesive and abrasive wear for the generation of wear particles. There are other cases of wear where a critical number of repeated contacts are essential for the generation of wear particles. Wear generated after such contact cycles is called "fatigue wear". When the number of contact cycles is high, the high-cycle fatigue mechanism is expected to be the wear mechanism. When it is low, the low-cycle fatigue mechanism is expected.

To better guide the choice of materials in the field, where surface interactions interferes with fatigue wear, it is necessary to understand the mechanisms and processes that govern the wear by contact fatigue with or without sliding. Detailed explanations can be found elsewhere [73, 74]. Nevertheless, in the following, a summary of this material is recalled concisely.

This form of contact fatigue-induced wear is often observed in systems where cyclic contact stresses (e.g., loaded tools) take place, but in most cases during sliding or rolling contacts.

In loaded mechanical parts, the contact surface undergoes compression stresses and shear stresses are developed beneath the surface. The repeated loading and unloading cycles to which the materials are exposed may induce the formation of surface and/or sub-surface nano- and microcracks, at critical zones where, for example, imperfections, inclusions or second phases are located. This eventually will result in the growth of fatigue cracks as further load stresses become apparent, and ultimately driving to the breaking-up of the surface with the formation and detachment of large fragments (i.e. sheet-like particles or spalling process), leaving large pits in the surface. This mechanism can be predominant, for example, in periodically loaded dies and tools, in roller bearings or in pumps that are exposed to cavitation.

Under tribological/tribocorrosion conditions, the location of maximum shear stress moves towards the near-surface during tangential sliding and micro-cracking occurs at the locations of surface defects. The transition from sliding wear to fatigue wear is evidenced by the initiation and propagation of one or more surface micro-cracks into the bulk material. The analysis of crack initiation in sliding (e.g., fretting) is far more challenging than that of propagation. A wide range of factors affects the number of cycles required for damage to accumulate but they can be conveniently be grouped into three categories [74]: (a) Material factor (grain size, morphology, fracture toughness, flow stress, etc.), (b) Environmental factors (temperature, humidity, etc.), and (c) Mechanical factors (surface tractions, slip amplitude, frictional dissipated energy etc.). A full understanding of the problem will only be gained when all these factors are brought under consideration. However, the range of expertise necessary to master these three areas is considerable. At present, few effects are particularly well understood (*viz.* contact stress field and surface slip amplitude) and approaches are developed accordingly. Two main approaches are available to develop an understanding of crack formation, either a "bulk" or macroscopic approach or a "local" or micromechanics approach. The actual crack initiation process occurring on the scale of a few grains in non-homogenous, anisotropic material riddled with defects, inclusions, and imperfections. There is some discrepancy in the literature as to what fraction of the total fatigue life is consumed by the crack initiation process.

Another challenge under hot working conditions has caught attention in mechanical loaded contacts: an unexpected heat may induce surface or volume expansion and the development of stresses between the surface and the bulk material. Between two loaded contact events, the surface cooling once more generates stresses, and "thermal fatigue" may occur resulting in a network of cracks resembling a mosaic.

Generally, contact fatigue can be mitigated by all usual measures that reduce the susceptibility to cracks, i.e., a high strength to impede crack initiation and a high toughness to retard crack growth (*cfr.* **Figure 1** in Section 2).

The relationship between these two material properties, namely toughness and strength, have been introduced in Section 2 (**Figure 1**). It is shown that, generally, ductile metals exhibit virtually the greatest fracture toughness; however, they may display weak strength. Metallic glasses, on the other hand, often have toughness and strength that lie between brittle ceramics and marginally tough materials. Nevertheless, following the development of very recent bulk metallic glasses, the introduction of newly metal glass composites on the market has led to overcome the early success of metals by acquiring a very high resistance to deformation, and toughness, which imparts them a good resistance to cyclic stresses and solicitations. A strong interface between the glassy matrix and second phase particles is to be guaranteed. Therefore, these glassy composites can substitute metals in application areas not achieved yet so far. Among these novel alloys, there are Pd-based glasses, and ductile-phase-reinforced metallic glass composites [2].

8. Future trends

All the metallic glass materials described in this chapter are subject to further development. Therefore, the attempt to sketch out all possible developments does not seem appropriate. In the design of materials, however, there is a general tendency to form graduated structures and multiphase models, i.e., materials that exhibit a property gradient from the surface to the core, and which can conceive a composite pattern.

With a better understanding of the mechanisms of tribocorrosion, knowledge-based development can lead to new microstructures capable of counteracting specific mechanochemical loads.

Author details

Abdenacer Berradja

Address all correspondence to: a.berradja@gmail.com

MTM Department, K.U. Leuven, Leuven, Belgium

References

[1] Schweiter PA. Metallic Materials: Physical, Mechanical, and Corrosion Properties. Boca Raton: CRC Press; 2003. ISBN: 9780824708788

[2] Demetriou M, Launey ME, Garrett G, Schramm JP, Hofmann DC, Johnson WL, Ritchie RO. A damage-tolerant glass. Nature Materials. 2011;**10**(2):123-128

[3] Irwin G. Analysis of stresses and strains near the end of a crack traversing a plate. Journal of Applied Mechanics. 1957;**24**:361-364

[4] Liang JZ, Li RKY. Rubber toughening in polypropylene – A review. Journal of Applied Polymer Science. 2000;**77**:409

[5] Fan C, Qiao D, Wilson T, Choo H, Liaw PK. As-cast Zr–Ni–Cu–Al–Nb bulk metallic glasses containing nanocrystalline particles with ductility. Materials Science and Engineering: A. 2006;**431**(1):158-165

[6] Fu H, Zhang H, Wang H, Zhang Q, Hu Z. Synthesis and mechanical properties of Cu-based bulk metallic glass composites containing in-situ TiC particles. Scripta Materialia. 2005;**52**(7):669-673

[7] Stachowiak GW, Batchelor AW. Engineering tribology. Tribology Series. 1993;**24**:872. ISBN: 978-0-444-89235-5

[8] Landolt D, Mischler S. Tribocorrosion of Passive Metals and Coatings. Cambridge UK: Woodhead Publishing Limited; 2011. ISBN: 978-1-84569-966-6

[9] Landolt D, editor. Corrosion and Surface Chemistry of Metals, (in Chap 10). New York: EFPL Press; 2007. pp. 415-460

[10] Celis JP, Ponthiaux P, editors. Testing Tribocorrosion of Passivating Materials Supporting Research and Industrial Innovation, Handbook EFC. (n° 62). New York: Maney Publishing, CRC Press Francis & Taylor Group; 2012. ISBN: 9781907625202

[11] Waterhouse RB. Fretting Corrosion. Oxford: Pergamon; 1972. pp. 182-196

[12] Landolt D, Mischler S, Stemp M. Electrochemical methods in tribocorrosion: A critical appraisal. Electrochimica Acta. 2001;**46**(24-25):3913-3929

[13] Ponthiaux P, Wenger F, Celis JP. Tribocorrosion: Material behavior under combined conditions of corrosion and mechanical loading (Chap 4). In: Shih H, editor. Corrosion Resistance. London, UK: Intech Open; 2012. ISBN: 978-953-51-0467-4

[14] Mischler S. Triboelectrochemical techniques and interpretation methods in tribocorrosion: A comparative evaluation. Tribology International. 2008;**41**:573-583

[15] Jiang XX, Li SZ, Tao DD, Yang JX. Accelerative effect of wear on corrosion of high-alloy stainless steel. Corrosion. 1993;**49**:836-841

[16] Garcia I, Drees D, Celis JP. Corrosion-wear of passivating materials in sliding contacts based on a concept of active wear track area. Wear. 2001;**249**:452-460

[17] Peter WH, Buchanan RA, Liu CT, Liaw PK, Morrison ML, Carmichael Jr CA, Wright JL. Localized corrosion behavior of a zirconium-based bulk metallic glass relative to its crystalline state. Intermetallics. 2002;**10**(11-12):1157-1162

[18] Qin CL, Zhang W, Asami K, Ohtsu N, Inoue A. Glass formation, corrosion behavior and mechanical properties of bulk glassy Cu–Hf–Ti–Nb alloys. Acta Materialia. 2005;**53**(14):3903-3911

[19] Pang SJ, Zhang T, Asami K, Inoue A. Synthesis of Fe–Cr–Mo–C–B–P bulk metallic glasses with high corrosion resistance. Acta Materialia. 2002;**50**(3):489-497

[20] Morrison ML, Buchanan RA, Peker A, Peter WH, Horton JA, Liaw PK. Cyclic-anodic-polarization studies of a $Zr_{41.2}Ti_{13.8}Ni_{10}Cu_{12.5}Be_{22.5}$. Intermetallics. 2004;**12**(10-11):1177-1181

[21] Pang SJ, Zhang T, Asami K, Inoue A. Formation, corrosion behavior, and mechanical properties of bulk glassy Zr–Al–Co–Nb alloys. Journal of Materials Research. 2003;**18**(7):1652-1658

[22] ASTM G61-86. Standard test method for conducting cyclic potentiodynamic polarization measurements for localized corrosion susceptibility of iron-, nickel-, or cobalt-based alloys. West Conshohocken, PA, USA: ASTM International; 2003

[23] ASTM Standard G3. Standard practice for conventions applicable to electrochemical measurements in corrosion testing. In: Annual Book of ASTM Standards. Vol. 03.02. Wear and Erosion; Metal Corrosion. West Conshohocken, PA: ASTM; 2001

[24] Yang L, editor. Techniques for Corrosion Monitoring. Series in Metals and Surface Engineering. 1st ed. UK: Woodhead Pub; 2008. ISBN: 1-84569-187-3

[25] Diomidis N, Celis JP, Ponthiaux P, Wenger F. A methodology for the assessment of the tribocorrosion of passivating metallic materials. Lubrication Science Journal. 2009;**21**(2):53-67

[26] Archard JF. Contact and rubbing of flat surface. Journal of Applied Mechanics and Technical Physics. 1953;**24**(8):981-988

[27] Rabinowicz E. Friction and Wear of Materials, 2nd ed., USA: Wiley Intersc. Pub., John Wiley & Sons, Inc.; 1995. ISSN 0021-8936

[28] Moore MA. Abrasive wear. In: Rigney DA, editor. Fundamentals of Friction and Wear. Metals Park, OH: American Society of Metals; 1980. p. 73

[29] Quinn TFJ. Review of oxidational wear, part I. Tribology International. 1983;**16**(5):257-270

[30] Welsh NC. The dry wear of steel. Part I. Philosophical Transactions of The Royal Society A. 1965;**257**:31-50

[31] Czichos H, Beeker S, Lexow J. International multilaboratory sliding wear tests with ceramics and steel (VAMAS 2nd round robin). Wear. 1989;**135**:171

[32] Berradja A, Bratu F, Benea L, Willems G, Celis JP. Effect of sliding wear on tribocorrosion behaviour of stainless steels in a Ringer's solution. Wear. 2006;**261**:987-993

[33] Mohrbacher H, Celis JP, Roos JR. Laboratory testing of displacement and load induced fretting. Tribology International. 1995;**28**(5):269-278

[34] Huq Z, Celis J-P. Expressing wear rate in sliding contacts based on dissipated energy. Wear. 2002;**252**(5-6):375-383

[35] Fouvry S, Kapsa P, Vincent L. Quantification of fretting damage. Wear. 1996;**200**(1-2): 186-205

[36] Matveevsky RM. Tansactions of ASTM. 1965;**87**:754

[37] Madsen BW. Measurement of erosion-corrosion synergism with a slurry wear test apparatus. Wear. 1988;**123**:127-142

[38] Watson SW, Friedersdorf FJ, Madsen BW, Cramer SD. Methods of measuring wear-corrosion synergism. Wear. 1995;**181-183**:476-484

[39] ASTM Standard. G119, Standard guide for determining amount of synergism between wear and corrosion. In: Annual Book of ASTM Standards. Vol. 03.02. Wear and Erosion, Metal Corrosion. West Conshohocken, PA: ASTM; 2001

[40] Mischler S, Debaud S, Landolt D. Wear-accelerated corrosion of passive metals in tribocorrosion systems. Journal of the Electrochemical Society. 1998;**145**:750-758

[41] Stachowiak GW, Batchelor AW, editors. Corrosive and oxidative wear (Chap. 13). In: Engineering Tribology book, Tribology Series. 1993;**24**:637-656. ISBN: 978-0-444-89235-5

[42] Zambelli G, Vincent L. Matériaux et Contact, Une Approche Tribologique. Lausanne: Press Polytechniques et Universitaires Romandes; 1998

[43] Ponthiaux P, Wenger F, Drees D, Celis J-P. Electrochemical techniques for studying tribocorrosion processes. Wear. 2004;**256**:459-468

[44] Shih H, editor. Corrosion Resistance. London, UK: Intech Open; 2012. ISBN: 978-953-51-0467-4

[45] Sedriks AJ. Corrosion of Stainless Steels. Princeton, NJ: The Electrochemical Soc.; 1979

[46] Fehlner FP. In: Frankenthal RP, Kruger J, editors. Passivity of Metals. Pennington, NJ: The Electrochemical Society; 1978. p. 181

[47] Wang Y, Jiang SL, Zheng YG, Ke W, Sun WH, Wang JQ. Electrochemical behaviour of Fe-based metallic glasses in acidic and neutral solutions. Corrosion Science. 2012;**63**:159-173

[48] Revesz AG, Kruger J. In: Frankenthal RP, Kruger J, editors. Passivity of Metals. Pennington, NJ: The Electrochemical Society; 1978. p. 137

[49] Macdonald DD, Urquidi-Macdonald M. Fundamental aspects in the design of passive alloys. In: Marcus P, Barouxand B, Keddam M, editors. EFC Pub. (n°12), Modifications of Passive Films, Papers of European Symposium on Modifications of Passive Films, Paris, France, 15-17 February 1993. Book Number 577. London, UK: Institute of Materials; 1994

[50] Lin LF, Chao CY, Macdonald DD. A Point Defect Model for Anodic Passive Films, II. Chemical Breakdown and Pit Initiation. Journal of Electrochemical Society. 1981;**128**(6):1194

[51] Hoar TP. Palladium Medal Address. Journal of Electrochemical Society. 1970;**17C**:117

[52] Sato N. A theory for breakdown of anodic oxide films on metals. Electrochimica Acta. 1971;**16**(10):1683-1692

[53] McBee CL, Kruger J. Nature of passive films on iron-chromium alloys. Electrochimica Acta. 1972;**17**(8):1337-1341

[54] Ryan MP, Newman RC, Thompson GE. Atomically Resolved STM of Oxide Film Structures on Fe-Cr Alloys during Passivation in Sulfuric Acid Solution. Journal of Electrochemical Society. 1994;**141**(12):L164-L165

[55] Meng HC, Ludema KC. Wear models and predictive equations: Their form and content. Wear. 1995;**181-183**:443-457

[56] Burwell JT. Survey of possible wear mechanisms. Wear. 1957;**1**:119-141

[57] ASTM G40-17. Standard Terminology Relating to Wear and Erosion, ASTM International, West Conshohocken, PA, 2017

[58] ASTM NACE/ASTMG193-12d. Standard Terminology and Acronyms Relating to Corrosion. West Conshohocken, PA: ASTM International; 2012

[59] Bhansali KJ. Wear coefficients of hard-surfacing materials. In: Peterson MB, Winer WO, editors. Wear Control Handbook. New York: ASME Publisher; 1980. pp. 373-383. ISBN: 9999999999

[60] Lancaster JK. Wear Mechanisms of Metals and Polymers. Transactions of the Institute of Metal Finishing. 1978;**56**(1):145

[61] Rabinowicz E. Wear coefficients – metals. In: Peterson MB, Winer WO, editors. Wear Control Handbook. New York: ASME Publisher; 1980. pp. 475. ISBN: 9999999999

[62] Lim SC, Ashby MF. Wear-mechanism maps. Acta Metallurgica. 1987;**35**(1):1-24

[63] Chiou YC, Kato K, Kayaba T. Effect of normal stiffness in loading system on wear of carbon steel – Part 1: Severe-mild wear transition. ASME, Journal of Tribology. 1985;**107**: 491-495

[64] Cho SJ, Hockey BJ, Lawn BR. Grain-size and R-curve effects in the abrasive wear of alumina. Journal of the American Ceramic Society. 1989;**72**(7):1949-1952

[65] Richardson TJA, editor. Shreir's Corrosion. Elsevier Science Pub. 1st ed. December 29, 2009. ISBN: 978-0-444-52787-5

[66] Vesselin D, Komatsu TJ. Interionic Interactions, Electronic Polarizability and Optical Basicity of Oxide Glasses. Journal of Ceramic Society of Japan. 2000;**108**:330-338

[67] Prakash B, Celis JP. The lubricity of oxides revised based on a polarisability approach. Tribology Letters. 2007;**27**(1):105-112

[68] Hutchings IM. Tribology: Friction and Wear of Engineering Materials. London UK: Edward Arnold; 1992. ISBN: 034056184X, 9780340561843

[69] Ludema KC. Scuffing, Run-in and the Function of Surface Films, Particularly Oxides. Review Paper for Interdisciplinary Collaboration in Tribological Project. USA: NASA-Lewis; 1981

[70] Czichos H. Tribology – A Systems Approach to the Science and Technology of Friction, Lubrication and Wear. Amsterdam: Elsevier Science Publishing Company; 1978

[71] Ashby MF. Materials Selection in Mechanical Design. 4th ed. Oxford: Butterworth-Heinemann; 2011. p. 90. ISBN: 978-1-85617-663-7

[72] Axén N, Zum Gahr K. Wear of TaC and TiC steel composite hard facings by soft and hard abrasives. Materialwissenschaft und Werkstofftechnik. 1992;**23**(10):360-367

[73] Suh NP. An overview of the delamination theory of wear. Wear. 1977;**44**(1):1-16

[74] Suresh S. Fatigue of Materials. 2nd ed. Oxford UK: Cambridge Univ. Press; 1998. ISBN: 9780521578479

Chapter 6

A Tribo-Electrochemical Investigation of Degradation Processes in Metallic Glasses

Abdenacer Berradja

Additional information is available at the end of the chapter

http://dx.doi.org/10.5772/intechopen.79387

Abstract

Metallic glasses are relatively new materials with a large potential for applications in various technical and biomedical fields. However, for efficient use of these novel materials with an interesting combination of properties, it is necessary to fully characterize them for their mechanical and electrochemical properties. Studies on the effects of chemical parameters (pH, temperature, concentration of reagent) and tribological parameters (load, sliding speed, counterbody, contact configuration) on the kinetics of the reaction (i.e., the material removal rate) supply information on the dominant mechanisms governing the tribo-electrochemical behavior of metallic glasses. Although considerable efforts have been made to characterize their mechanical, corrosion, and magnetic properties, the study of their tribocorrosion patterns is in a rather unsatisfactory state, and very limited information is available. It is the purpose of this chapter to provide an overview of basic information on the tribo-electrochemical properties of most metallic glasses. This becomes crucial when such materials are to be considered in systems where solid surfaces are prone to mechano-chemical transformation processes.

Keywords: bulk metallic glasses, friction, wear-corrosion, synergy, tribocorrosion, tribo-electrochemistry

1. Introduction

Since the first discovery of amorphous alloys in the twentieth century, by Kramer [1] in the 1930s through vapor deposition, succeeded by Brenner et al. [2] in the 1950s through electrodeposition, and then by Klement et al. [3] in the 1960s through rapidly quenching casting processes (related to Au70Si30 glassy alloy system), several bulk glass-forming systems have emerged. Further innovation in this novel area of materials science has accredited a potential market for commercial applications of metallic glasses. As a result, increased research on various tribological and corrosion properties of BMGs has occurred. The first published information on the corrosion properties of metallic glasses appeared in 1974 (it concerned the Fe-Cr-P-C alloy system) [4]. Since then, the corrosion behavior of all classes of metallic glasses and/or amorphous nanocrystalline alloys has been of great concern.

2. Corrosion resistance of bulk metallic glasses

The corrosion resistance is a critical factor for considering bulk metallic glasses (BMGs) used in hostile or chemical environments. However, there has been a growing interest in the corrosion behavior of these amorphous alloys.

At first presumption, many metallic glasses exhibit extremely good corrosion resistance, but there is a controversy over the exact mechanism(s) responsible for such improvement with respect to their crystalline counterparts. There are several possibilities for explaining the difference in corrosion behavior between amorphous and crystalline metals. The good corrosion resistance of single-phase glasses is often attributed to structure, composition, as well as structural and compositional homogeneity. Both the chemical homogeneity and the absence of microstructure in these amorphous alloys are most likely source of their superiority over their crystalline candidates. It has commonly been assumed that the very short solid-state diffusion time caused by the rapid cooling rates (as high as 10^6 K s^{-1}) required to produce the amorphous alloys, prevents the chemical heterogeneity and crystallization (but not always) [5].

When diffusion in the solid-state is kept faster and faster enough, the yielded homogeneous amorphous alloys should lack grain boundaries, dislocations, second-phase particles, segregates and other structural defects. These structural impurities are often present and commonly the culprits behind the localized corrosion (e.g., pitting, galvanic corrosion, etc.) observed in their crystalline counterpart alloys exposed to an aggressive environment. Corrosion readily occurs preferentially at such defect sites; therefore, metallic glasses can be expected to exhibit better corrosion resistance than crystalline alloys. Nevertheless, it has been shown that this statement does not always apply to BMGs [6]. Indeed, second phases (crystalline inclusions) were observed in some BMGs even after treatment. This was attributed to heterogeneous nucleation reactions caused by impurities in the melt that remained on the metal surface of the finished good [7].

Under normal working conditions, cathodes and anodes, possibly small but sufficiently large to cause localized corrosion, can be formed on the same metal glassy surface. Corrosion at the origin of these local crystalline inclusions will be greatly amplified upon conditions and can cause corrosion pits. This is one of the reasons why the persistent quest for BMG systems with increased glass-forming ability (i.e., decreased crystalline inclusions, see previous chapter) has provided the field with an influx of new alloying BMGs and compositions.

The ability of metals for alloying has provided the key to achieve enhanced properties, but certain elements are more common than others are. It is mainly the combined effect of the alloying elements, the test environment and, to some extent, earlier specific treatment conditions, which determines the electrochemical properties profile of a number of metallic glasses. The addition of simple metals (SM), transition metals (early and late TM), and rare earth metals (REM) to the base metal was, inter alia, the principal strategy used to produce a selection of different metallic material glasses with a broad range of alloy compositions. According to IUPAC definition, a transition metal (TM) is defined as "a chemical element whose atoms have an incomplete *d* electronic sub-shell, or which can form cations whose electronic sub-shell *d* is incomplete."

The early *versus* late transition metals differ generally in their oxidation states (each metal has different possible oxidation states). Electrons have a stronger attraction to the protons in the late transition metals (LTM), so the (LTM) form bonds that are harder to break.

Metallic glass alloys can be grouped into two major distinctive categories with intrinsically different corrosion behaviors, namely:

- The first group includes the TM-metalloid alloy systems. These alloys are usually the early transition metals (ETM): iron-, copper-, nickel-, cobalt-, zinc-, etc., and late transition metals (LTM): chromium-, zirconium-, titanium-, manganese-, lanthanum-, niobium-, scandium-, yttrium-, etc., base systems, and may normally contain approximately 20 at.% P, B, Si, and/or C as the metalloid component.
- The second class consists of binary, ternary or quaternary alloy system such as TM-SM, SM-TM-RE, and assorted SM-TM-X alloys. These alloys can contain a RE and/or TM and SM elements (such as aluminum, magnesium, beryllium, etc.).

2.1. Corrosion and associated mechanisms in the case of the transition metal-metal binary alloys

2.1.1. Fe-based BMG materials

One of the earliest studied systems in metallic glasses (MG), ranging from alloy design to material properties, is that of Fe-base alloys. These alloys can be formed as binary systems, such as Fe-B, or they may be much more complex, such as Fe-Ni-P-B quaternary systems, and even more intricate multiple systems. It has been suggested that the excellent resistance of certain iron metallic glass alloys to both uniform and localized corrosion results from their enhanced chemical reactivity relative to conventional stainless alloys [8]. The resistance to corrosion may be a result of the formation of a kind of passive film consisting of metallic and metalloid elements capable to strengthen the metallic glass surface against any specific chemical attack and under aggressive conditions (*cfr.* previous chapter).

In addition to the base metal, metallic glass alloys often contain appreciable concentrations of oxide film builder alloying elements to promote passivity, such as Cr, Ni, P, B as in the case of Fe-Ni-Cr-P-B system. They derive their corrosion resistance from similar shielding process as that of crystalline alloys, namely the development of a passive protective film (oxide scales). The significant difference between the corrosion-resistant metallic glass alloys and their crystalline counterparts, such as stainless steels, lies in the fact that the level of chromium necessary to promote passivity can be substantially less in the metallic glass alloys.

2.1.1.1. *Effect of Cr content*

One of the pioneer corrosion resistance studies on Fe-based amorphous alloys with P and C contents in acid chloride-based solutions was performed by Naka et al. [9]. Both uniform and localized corrosion resistances of these glassy alloys were mainly attributed to the growth of protective oxide films when appropriate additions of C and P are used [10]. A comparison was made between the corrosion rates of crystalline Fe-Cr alloys and amorphous Fe-Cr-P-C alloys as a function of chromium concentration as shown in **Figure 1**. At low chromium atomic content levels (<4 at.%), the amorphous alloy corrodes at a higher rate than the crystalline material. However, at even higher chromium levels (4 at.%), there is a significant decrease in the corrosion rate of the MG alloy, whereas the crystalline material remains unchanged. At more elevated and intermediate content Cr level (8 at.%), no corrosion was detected on the MG alloy by weight loss experiments after immersion for 168 h. Surprisingly, it was found that the concentration of HCl electrolyte, which has a significant effect on the corrosion behavior of the crystalline alloys, had no effect on the corrosion of the Fe-Cr-P-C or Fe-Ni-Cr-P-C base MG alloy systems, and which exhibited no weight loss after exposure for a period of 168 h [11].

Chromium was shown to be very effective to confer pitting resistance, such as for the metallic glass alloys Fe-CrxB13C7 and Fe-CrxB13Si7 in 3% NaCl [8]. With chromium level contents of 2 and 5 at.%, both MG alloy types suffered pitting at potentials more anodic than the free corrosion potential of approximately 0.6 V/SCE. The addition of 8 at.% Cr extended the pitting resistance nearly to 1 V/SCE, which represents an extreme condition of competition for alloys containing such a low level of chromium. By contrast, type 304 stainless steel contains

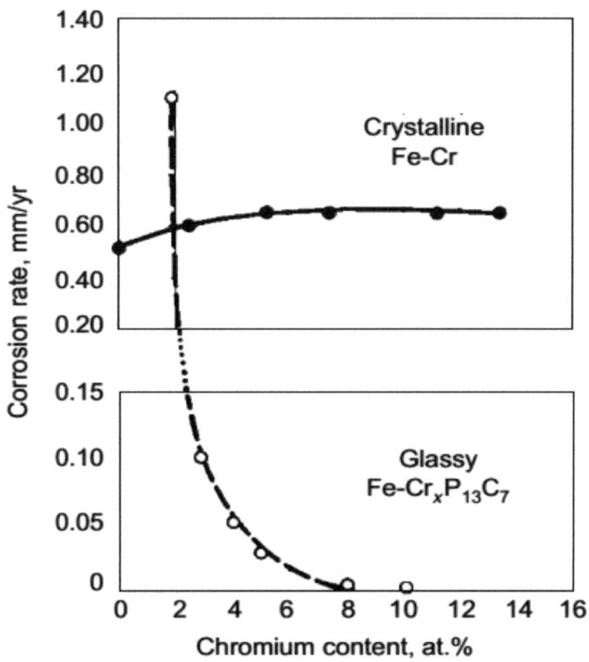

Figure 1. Comparison of the corrosion rates of glassy FeCrXP13C7 alloys and crystalline iron-chromium alloys in 1 N NaCl solution at 30°C. Reproduced from [11] with permission from Elsevier Science.

approximately 18 wt.% Cr, yet its pitting potential is several hundred millivolts less positive than that of the foregoing metallic glass alloys.

In another similar study, it has been shown that an increase of the Cr content from 0 up to 16 at.% in a series of Fe-Ni-Cr-P-B MG alloy filaments systems facilitated passivation in acidic electrolyte (1 N NaCl), and no pitting was observed on any of such alloy system polarized below the transpassive potential region [12]. However, polarization at transpassive potentials caused numerous pits to form that penetrated the MG alloy filament surface. Pits were found to be noncrystallographic in shape.

Pang et al. [13] investigated the corrosion behavior of bulk Fe50-XCr16Mo16C18BX alloys (X = 4, 6, and 8 at.%) in HCl solutions. The increase of B level content up to 8 at.% led to a decrease of the passive current density and to the reduction of the corrosion rate as well. Likewise, in other similar corrosion investigations reported elsewhere [13, 14], the Fe43Cr16Mo16C10B5P10 BMG exhibited a lower passive current density than that of Fe43Cr16Mo16C15B10 BMG in a 1 M HCl electrolyte. However, in a 6 M HCl solution, and upon anodic polarization, the current density abruptly increases for the Fe43Cr16Mo16C15B10 BMG, whereas the Fe43Cr16Mo16C10B5P10 BMG displayed a larger region of passive range with a passive current density of 100 mA m^{-2}, which makes it more effective in increasing the corrosion resistance compared to the former. A summary of these findings, which shows a decrease in the passive current density, and the expansion of the passive region as a function of B and P additions to Fe-based BMGs tested either in a 1 M HCl or in a 6 M HCl solution, is displayed in **Figure 2**.

2.1.1.2. Effect of Mo content

Molybdenum benefits the pitting resistance of metallic glass alloys and crystalline steels. The addition of Mo to glassy Fe-MoxP13C7 alloys suppressed pitting, decreased the critical current density for passivation and the passive current density [15]. As little as 4 at.% Mo prevented pitting corrosion in 1 N HCl; likewise, low additions of Mo were shown to be more effective in decreasing the corrosion rates than Cr. Molybdenum has been shown to facilitate the formation of a passive film of hydrated chromium or iron oxyhydroxide through its enrichment in the corrosion product layer during active dissolution [16]. The enrichment helps for the accumulation of passivating species in the film by lowering the dissolution rate of the species; the Mo-rich product subsequently dissolves, leaving little Mo behind in the film. However, in another case study [17], the increase of Mo content level up to 22.5 at.% was shown to be detrimental to the corrosion resistance of Fe-based BMG, more likely by increasing the risk of formation of secondary phases in the ferritic matrix zone.

2.1.1.3. Effect of other elements content (Ti, Zr, V, Nb, W, Mn, Co, Cu, Ru, Rh, Pd, Pt)

The effect of other alloying elements to Fe-based glass alloy on its corrosion resistance has been addressed in one study [18]. Hence, titanium, zirconium, vanadium, niobium, chromium, molybdenum, tungsten, manganese, cobalt, nickel, copper, ruthenium, rhodium, palladium, and platinum were all added to the Fe-X-P13C7 glassy alloy [18]. All elements, except manganese, decreased the corrosion rate of the iron glassy alloy in H_2SO_4, HCl, HNO_3, and

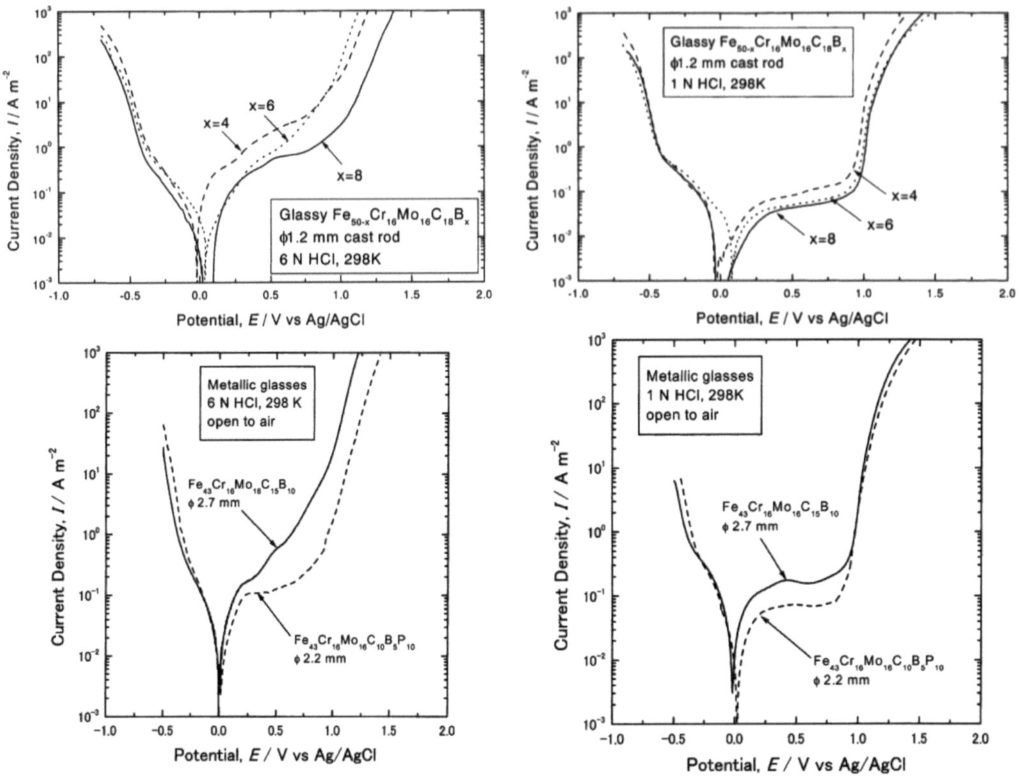

Figure 2. Potentiodynamic polarization curves of the cast glassy Fe-based BMGs (rods with a diameter of 1.2 mm) in 1 and 6 M HCl open to air at 298 K. Reproduced from [13, 29] with permission from Elsevier Science.

NaCl solutions. Although the base alloy, Fe-P13C7, did not passivate; additions of any of the foregoing elements at levels from 0.5 up to 40 at.% enabled passivation to occur during anodic polarization in 0.1 N H_2SO_4. Chromium was the most efficient, still, molybdenum, and titanium were very beneficial. No pitting was observed in 3% NaCl for passivated alloys. The alloys that did not passivate, such as Fe-Co-P13C7, did not pit, but rather they dissolved uniformly.

Fe-W resisted to pitting corrosion up to 2.5 V/SCE in both acidic and neutral chloride solutions (pH 1 and 7, respectively) [19]. Addition of tungsten to Fe-WxP13C7 has the effect of increasing the critical pitting potential, E_{crit}, to a level above 2 V/SCE for x = 6 at.%, but when x = 10 at.% of W was added it caused transpassive dissolution at 1 V/SCE of the MG alloy [19].

2.1.2. Ni-based BMG materials

Generally, Ni-based metallic glass systems exhibit a good resistance to uniform and localized corrosion. A number of investigations on the electrochemical characteristics of Ni-based amorphous alloys have been performed on ribbons [20–22] due to the struggle in producing amorphous bulk samples (i.e., having thickness > 1.5 mm). The elemental constituents that have typically been used to ensure a good corrosion resistance were either additions of metalloids such as P [20, 22] or additions of metals such as Ta [20, 21] and Nb [20].

http://dx.doi.org/10.5772/intechopen.79387

Metallic glassy Ni-P has been recently investigated, which appeared to resist to chloride-induced corrosion [23]. In fact, its *E*-log (*i*) potentiodynamic behavior was basically identical whether in a chlorinated or chlorine-free environment. A form of chemical passivity has been proposed to explain its corrosion behavior. Passivation in the Ni-P system was mainly due to the formation of an ionic barrier layer rather than a classical passive oxide film. This barrier layer consists of hypophosphite ion adsorbed on the nickel phosphorous surface, with hydrogen/H_2O bonded outer layer. This barrier reaction layer inhibits the ion transport of water to the surface and thus prevents the hydration of nickel, which is the first step in the nickel dissolution process.

More recently, there have been growing efforts to decrease the additions of P due to its effective action on the loss of ductility of the Fe amorphous alloy [24]. Overall, the strategic approach that is found to be most effective in increasing the corrosion resistance depends on whether the solution is strongly oxidizing or not (e.g., 9 M HNO_3).

2.1.2.1. *Effect of elements content (Ti, Zr, Ta, Nb, Cr, P)*

There has been a consensus that additions of suitable metals to amorphous Ni-based alloys tend to increase their corrosion resistance. The electrochemical behavior of Ni-based amorphous alloys containing Ti, Ta, Zr, Nb, Cr, and/or P has been of a great concern for a number of investigators [20, 25–27]. One of the most significant studies was led by Shimamura et al. [20] who investigated the effect of P and other valve metals (e.g., Ta) on the corrosion properties of Ni-based amorphous ribbons immersed either in boiling 9 M HNO_3 solutions with and without any of Cr^{6+} ions content or in a boiling 6 M HCl electrolytes. Ta additions have been proven to be the most efficient in lowering the corrosion rates of Ni-based MG alloy. The addition of critical amounts of Ta resulted in undetectable corrosion rates ($<10^{-3}$ mm year^{-1}). For example, after being immersed in a boiling 9 N HNO_3 solution for 168 h, the corrosion rate of Ni60Ti40 was estimated to be close to 1 mm year^{-1}. After adding of 30 at.% Ta, however, the Ni60Ti10Ta30 MG alloy exhibited an immune response to corrosion for the same period of exposure, i.e., 168 h. Immunity was attributed to the formation of a kind of a protective layer. Although, the authors Shimamura et al. [20] claimed that amorphous Ni-Ta alloys required more than 35 at.% Ta in a boiling 6 M HCl solution to form a tantalum oxyhydroxide ($TaO_2[OH]$) shielding passive film.

Alternatively, the addition of a small amount of P to Ni-Ta glassy alloys has been proven to be effective in significantly reducing their corrosion rates. The corrosion rate of Ni70Ta30 in a boiling 6 M HCl solution was more than 10^4 times greater than that of Ni68Ta30P2 alloy when tested under similar conditions [20]. The authors believed that the addition of P promoted the growth of $TaO_2(OH)$ passive film by accelerating selective dissolution of elements unnecessary for the passive film formation [20]. However, when experiments were performed in solutions with a high oxidizing power, the authors found that the addition of P to Ni-Ta alloys was not necessary to promote the growth of the passive film. Interestingly, many research works have suggested that a Ta-enriched passive film would probably be one of the reasons for the high corrosion resistance of Ni-based amorphous alloys in aggressive solutions [20, 21].

Moreover, it has been proven that the addition of approximately of 7 at.% Cr was sufficient to prevent pitting corrosion of Ni-Cr-P-B alloy systems immersed in 10% $FeCl_3.H_2O$ at 30°C [27].

In another electrochemical study on the Ni-based MG alloy, Habazaki et al. [25] carried out potentiodynamic polarization tests on Ni75-XCrXTa5P16B4 BMGs (X = 5, 10, and 15 at.%) in a 6 M HCl solution in open air at 303 K. The passive current density was shown to decrease as the Cr (at.%) content increased in Ni60Cr15Ta5P16B4.

Not all metal additions can improve the corrosion resistance of Ni-based alloys in aggressive solutions. This was the case for Cobalt. Pang et al. [26] studied the anodic polarization behavior of Ni60-XCoXNb20Ti10Zr10 (X = 0, 5, and 20 at.%) in a 6 M HCl solution. The Co additions did not significantly alter the polarization behavior; however, the three tested compositions exhibited spontaneous passivation in the 6 M HCl solution and no pitting was experienced during the anodic polarization. Passive current densities for the three tested alloys were almost identical, that is, about 10^2 mA m^{-2}. The beneficial effects behind the modification of Ni-based amorphous alloys are summarized in **Figure 3**.

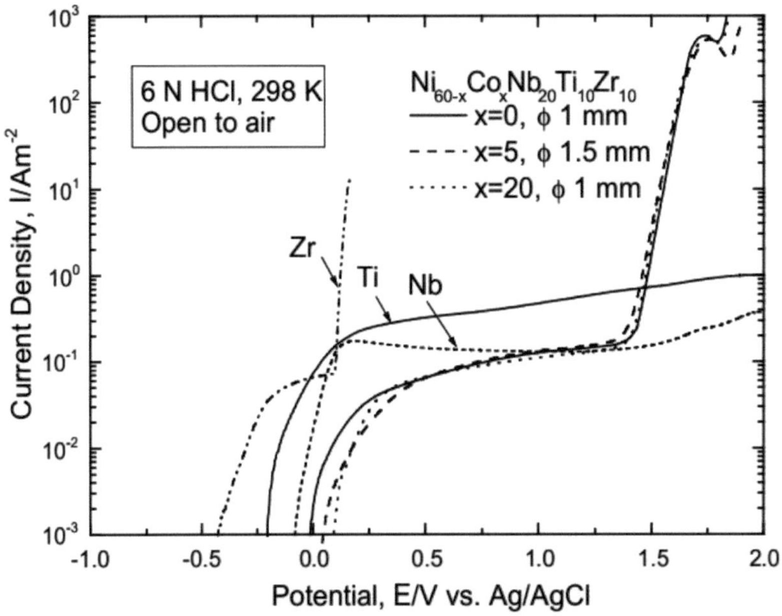

Figure 3. Anodic polarization curves of the bulk glassy Ni60-xCoxNb20Ti10Zr10 alloys with their critical diameters for glass formation and pure niobium, titanium and zirconium in 6 N HCl solution open to air at 298 K. Reproduced from [26] with permission from Elsevier science.

2.1.3. Cu-based BMG materials

Recently, the corrosion behavior of either Cu, Zr, or Cu-Zr glassy alloy systems has been investigated [28]. The potentiodynamic anodic polarization behavior of the alloy system exhibited the characteristics of two components, namely Cu, and Zr, while the corrosion resistance of the alloy was not greater than that of the more passive (and noble) metal of the alloy, *viz.* zirconium [28]. Thus, it can be deduced that the corrosion resistance in some transition metal-metal alloy systems could seemingly be due to the presence of a key passivation element in the solid substitute solution, and not to the glassy structure.

In line with previous conclusions, and in order to better understand the origin of corrosion resistance in MT-MT systems, another supported corrosion study was carried out on alloys of Cu-Zr and Cu-Ti systems in H_2SO_4, HCl, HNO_3, and NaOH solutions [15]. In all solutions except the alkaline one, that is, NaOH, both crystalline and metallic glass Cu-Ti alloys exhibited corrosion rates lower than those of pure Cu, and in all cases, the corrosion resistance of the metallic glass alloy was better than that of the crystalline alloy. The metallic glass alloys in such compositional systems are not unusually corrosion-resistant; in fact, neither the crystalline nor the glassy forms of the alloys were more corrosion resistant than pure Ti or pure Zr. This may suggest that the presence of passivation elements, such as Ti, or Zr promotes the corrosion resistance of the metallic glass alloys, and therefore it cannot barely be the result of the presence of the glassy or vitreous state. These results are consistent with previous findings in that the corrosion resistance could be determined by the behavior of the most corrosion-resistant component of the glassy alloy in the transition metal-metal systems.

The synthesis of new bulk metallic glass alloys has subsequently extended the field of corrosion study of many alloy systems, including BMGs. Thus, the development of Cu-based BMGs has been a relatively new occurrence. Lin and Johnson [30] have successfully synthesized Cu-Zr-Ti-Ni glassy alloys with thicknesses of at least 4 mm. Afterward, Inoue et al. [31] developed Cu-Zr-Ti amorphous alloys containing at least 50 at.% Cu with critical diameters between 4 and 5 mm. Among the Cu-Ti glassy alloys, the Cu-Zr-Ti [32] and Cu-Hf-Ti [31] BMGs earned more attention because of their excellent mechanical properties (*viz.* compressive fracture strengths of 2.06–2.15 GPa), which put forward the concept of their possible usage as engineering materials.

The chemical resistance of Cu-BMG systems has become more and more interesting. Unfortunately, the corrosion resistance of most amorphous Cu-based bulk alloys has not been as impressive as their mechanical properties. Nevertheless, the influence of the content of the composition on the corrosion behavior of Cu-based BMGs with a variety of additions of elements is still in a continuous evaluation phase. Many of such additions have shown to improve the corrosion resistance of Cu-based BMGs [33–37, 39]. Additions of small amounts of Nb [33, 35, 36, 39] led to increase the corrosion resistance of Cu-based BMGs. The alloy with other elements such as Cr [34], Ta [33], or Mo [33, 34, 37] has also proved to be effective in improving the electrochemical properties of the Cu-base BMG systems (e.g., corrosion and pitting potentials, etc.). This was the case of one of the comparative studies conducted by Inoue, and Qin et al. [33, 35, 36], who evaluated the effect of low additions of Nb, Mo, and Ta (to a Cu-Zr-Ti-X (X = Nb, Mo, Ta)) on the corrosion behavior of the Cu60Zr30Ti10 (at.%) BMGs exposed to solutions of 1 M HCl, 1 M HNO_3, 1 M NaOH, or 0.5 M (3%) NaCl.

The typical potentiodynamic polarization curves of these Cu-based BMGs are shown in **Figure 4**. The electrochemical behavior of Cu59.4Zr29.7Ti9.9Nb1, Cu59.4Zr29.7Ti9.9Mo1, and Cu59.4Zr29.7Ti9.9Ta1 demonstrated that the addition of Nb was the most effective element in lowering the corrosion rate of Cu-based systems in all the test solutions considered. An increase in the Nb content (up to 5 at.%) led to a decrease in the rate of corrosion in all the solutions tested. As can be seen in **Figure 4**, the additions of Nb to Cu-base alloy resulted into more positive values of the E_{cor}, an indication of the improvement of its nobility, thereby suggesting a better behavior of corrosion resistance. This trend was supported by the lower

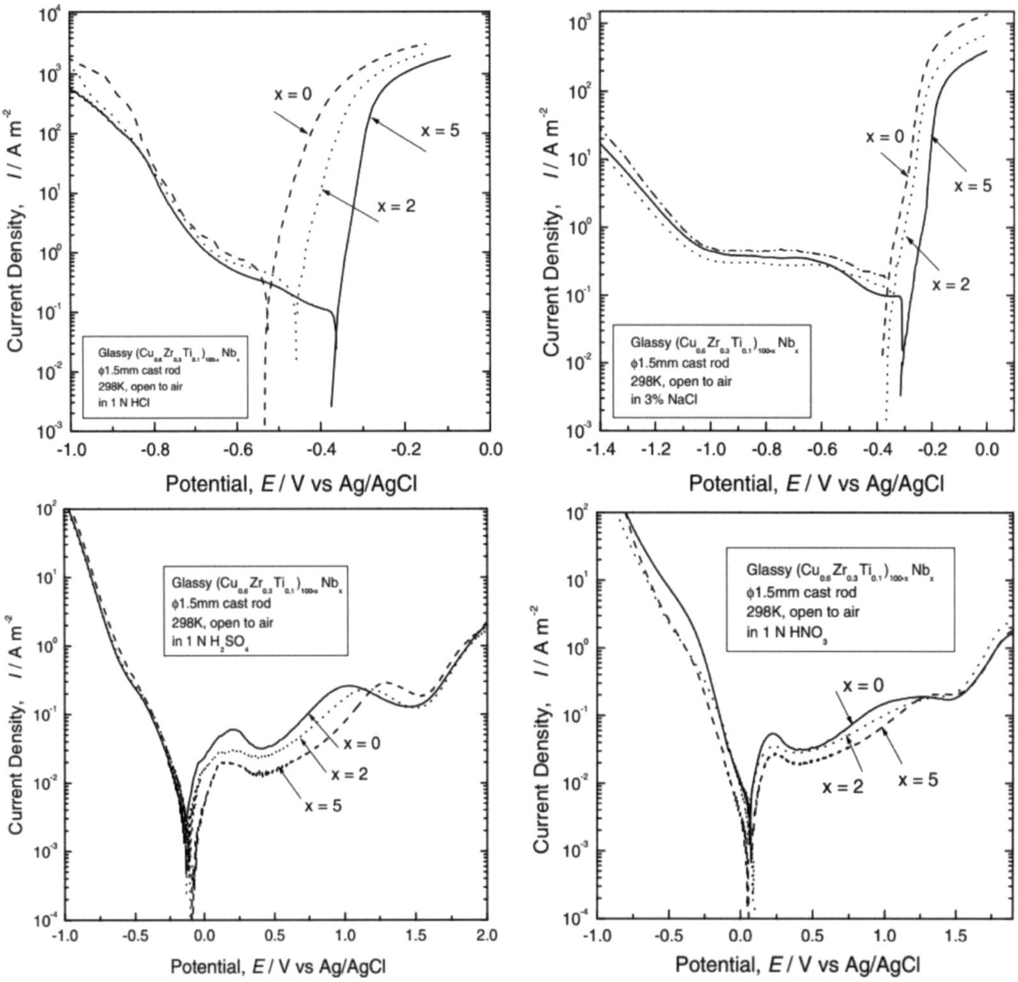

Figure 4. Potentiodynamic polarization curves of Cu-based BMGs (Cu0.6Zr0.3Ti0.1)100xNbx (x = 0, 2 and 6 at.%) alloys) in either 1 N HCl, 0.5 M (3%) NaCl, 1 N H_2SO_4 or 1 N HNO_3 open to air at 298 K. Reproduced from [35] with permission from Elsevier Science.

i_{cor}, and the larger E_{pit} values. Ta was not to, as effective as Nb, more likely due to its lower concentration (only 0.2 at.% Ta was added).

Liu and Liu [34] investigated the electrochemical behavior of Cu47Zr11Ti34Ni8 and (Cu47Zr11Ti34Ni8)99.5X0.5 (X = Cr, Mo, and W) BMGs in aqueous solutions of 0.5 M H_2SO_4 and 1 M NaOH. Additions of Cr, Mo, and W have led to the extension of the passive region, the lowering of passive current densities, and lowered the corrosion rates. The alloy with Mo addition, however, showed the most improved corrosion resistance in the two tested solutions, namely H_2SO_4 and NaOH. All Cu-based BMGs with Mo additions had passive films enriched in ZrO_2 and TiO_2 but depleted in Cu-oxides, which are less chemically stable and denser than ZrO_2 and TiO_2 [37]. It was believed that the addition of Mo was most efficient in improving the corrosion resistance of the alloy because its lower ionization energy compared to that of Cr and W, and leading to faster film formation [34, 37].

Attention was paid to the dependence of the corrosion behavior of Cu-based BMGs on the test environment [33, 35, 36, 38, 39]. Different electrolytes have been selected to study this dependency. In their corrosion study of Cu-Zr-Al-Nb BMGs in 1 M HCl, 0.5 M NaCl, or 0.5 M H_2SO_4, and regardless of the composition, Tam et al. [38] showed that the corrosion rate of Cu-Zr-Al-Nb BMGs was higher in the case of more aggressive solutions (i.e., 1 M HCl and 0.5 M NaCl) and lower in the less aggressive solution (0.5 M H_2SO_4). In both solutions containing chloride ions, the Cu-BMGs exhibited active behavior, demonstrating the deleterious effects of chloride ions. However, an active-passive behavior was found in the case where Cu-BMGs were exposed to H_2SO_4. Similarly, Qin et al. [39] showed that either Cu-Zr-Al BMG or Cu-Zr-Al-Nb BMG exhibited an active behavior in a 0.5 M NaCl and escorted with high corrosion rates, and an active-passive response accompanied with low rates of corrosion in 0.5 M H_2SO_4 solution.

2.1.4. *Zr-based BMG materials*

So far, there have been only a few attempts to study the electrochemical properties of Zr-based BMGs in recent years. The majority of these studies have been devoted to systems such as Zr-Ni-Cu-Al [40], Zr-Ti-Ni-Cu-Be [41], and Zr-Ti-Ni-Cu [30] alloys. These Zr-based BMG families are attractive due to their good glass-forming ability and excellent mechanical properties.

Since then, the procedure of adding noble elements (e.g., Nb, Pd, Ti, Ta) has enabled it possible to develop new compositions of MG capable of improving both the capacity of glass-forming ability and the resistance to uniform and localized corrosion [42–45]. Raju et al. [45] investigated the corrosion behavior of Zr-Cu-Al-Ni-X (X = Nb or Ti up to 5 at.%) BMGs in low alkaline sulfate (0.1 M Na_2SO_4), and chloride electrolytes (0.01 M NaCl). Increasing the Nb or Ti content resulted in a slight decrease in E_{cor} and an increase in the passive current density in the sulfate solution. On the contrary, such an increase in the contents of Nb and Ti leads to an increase in the values of η_{pit} and η_p in the NaCl solution, thus revealing the improvement of the alloy *vis-a-vis* its corrosion resistance. A typical comparative graph of the anodic polarization curves of these related Zr-based BMGs in 0.1 M Na_2SO_4 electrolytes (pH 8) is shown in **Figure 5**.

In a similar study, Asami et al. [42] investigated the electrochemical behavior of Zr60-XNbXAl10Ni10Cu20 (X = 0, 5, 10, 15, and 20 at.%) BMGs in 0.5 M H_2SO_4 and 1 M HCl solutions. An increase in the concentration of Nb resulted in an increase in the E_{pit}. The η_{pit}, however, did not always increase when the Nb content was increased. Substitution of 20 at.% of Nb for Zr resulted in a decrease in the rate of corrosion penetration (CPR) in the 1 M HCl solution from 100 down to 1 μm $year^{-1}$. The effectiveness of Hf addition (at.%) in improving the passivation ability of the Zr-Cu-Ni-Al BMG has been proven by Liu et al. [46]. Nevertheless, it was pointed out that the addition of noble metals does not always guarantee a better corrosion resistance of Zr-base BMGs.

Qin et al. [43] investigated the corrosion resistance of Zr-base bulk amorphous alloys with three different compositions, namely Zr55Al10Cu30Ni5-XPdX (with X = 0, 1, 3, and 5 at.%) in 0.6 M NaCl. Their findings have shown that the additions of Pd to the Zr-based MG lead to a decrease in the value of η_{pit}. Therefore, a change in the composition of Zr-based BMG does not always have a shielding effect for the MG alloy.

The effect of the environment is as challenging as that of alloying. It has been shown to play a significant role in the corrosion behavior of Zr-based bulk amorphous alloys. Gebert et al. [47]

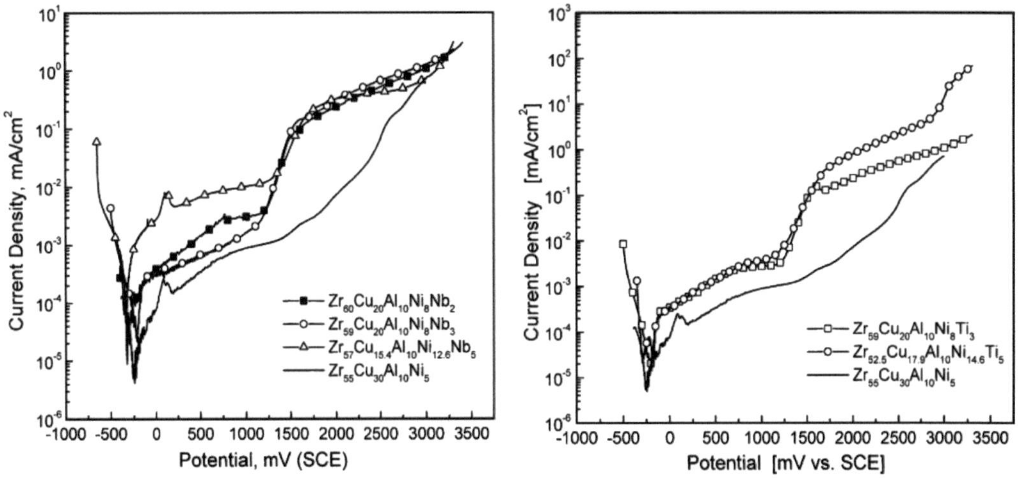

Figure 5. Anodic polarization curves of Zr-Cu-Al-Ni-x (x = Nb or Ti) BMG alloys in 0.1 M Na_2SO_4 solution (pH 8). Reproduced from [45] with permission from Elsevier Science.

showed that the Zr55Cu30Al10Ni5 exhibited an immune response to localized corrosion either in a 0.1 M Na_2SO_4 or in a 0.1 M NaOH solution over the entire potential scanning range (−1000 up to 2000 mV/SCE). However, the susceptibility to pitting corrosion was observed on the BMG surface during anodic polarization experiments for chloride concentrations as low as 10^{-3} M. The E_{pit} was shown to decrease as the chloride concentrations increased. However, this trend has been tempered by the anodic pre-growth of a passive film. Similarly, Mudali et al. [48] have found that an increase in the concentration of NaCl of 0.01–0.2 M, added to a 0.5 M H_2SO_4 electrolyte, significantly decreased the E_{pit} of the Zr-base BMG. Other studies [47, 49] further support this outcome. It has been shown that a decrease in localized corrosion resistance is appropriately associated with the exposure of the Zr-base BMG surfaces to solutions with increasing concentrations of chloride ions [47, 49]. In agreement with the foregoing conclusions, many researchers [42, 44, 48, 50, 56] have concluded that the majority of the degradation of a number of Zr-based BMG systems due to localized corrosion involved exposure to solutions containing chloride ions while their best performance was satisfied in H_2SO_4, Na_2SO_4, and NaOH solutions.

The effect of other factors, despite what has been mentioned above, such as the test temperature and passivation level, was found to affect the electrochemical properties of Zr-based BMGs. Gebert et al. [50] studied this effect on the corrosion behavior of Zr55Cu30Al10Ni5 BMG. The anodic polarization was carried out at 298, 423 and 523 K in a 0.001 M NaCl electrolyte on the pre-passivated Zr55Cu30Al10Ni5 samples and those without any specific treatment. For untreated and pre-passivated BMG samples, the E_{pit} decreases with increasing temperature. It was concluded that a decrease in temperature and prior passivation treatment promoted the tendency of the Zr55Cu30Al10Ni5 BMG to resist pitting in the chloride solution, which was consistent with the observations of many crystalline and amorphous metal systems.

2.2. Corrosion and associated mechanisms in the case of the binary, ternary, or quaternary amorphous alloy systems

2.2.1. *Other BMG-based system materials*

The majority of previous corrosion studies involved BMG systems based on Cu, Fe, Ni, and Zr, but there are other BMG systems, such as Ca- [51] Mg- [52], and Ti-based MG [53] still under investigation.

The selection of materials and design alloys are, inter alia, the major factors driving the global BMGs market. More new BMG systems will emerge and become commercially available in mass production in the near future as these amorphous alloys have many attractive properties for everyday life such as biomaterials, electronic devices, structures, and so on. This is the case for Ca- and Ti-based BMGs, which have shown great interest because of their potential applications as biomaterials. The Mg-based system is interesting for applications requiring high strength with lightweight materials [54].

The electrochemical properties of Ca-based BMGs (Ca65Mg15Zn20, Ca55Mg18Zn11Cu16, and Ca50Mg20Cu30) were investigated in a 0.05 M Na_2SO_4 electrolyte [51]. The Ca65Mg15Zn20 BMG experienced pitting at free corrosion conditions and had a CPR of 5691 μm year^{-1}. However, both Ca50Mg20Cu30 and Ca55Mg18Zn11Cu16 were slightly passivated at E_{cor} conditions and exhibited CPR values in the order of 1503 and 311 μm year^{-1} respectively.

The electrochemical behavior of the Ti43.3Zr21.7Ni7.5Be27.5 BMG immersed in a phosphate buffered saline (PBS) solution at 310 K was examined by Morrison et al. [53]. The Ti-base BMG exhibited a passive behavior at E_{cor} conditions but it showed a localized corrosion susceptibility at more increasing potentials. The η_{pit} value was about 589 mV/SCE. The Ti-base BMG alloy had a CPR value of about 2.9 μm year^{-1}. The authors concluded that the alloy resistance to localized corrosion in the PBS solution was equivalent to or greater than that of the 316 L stainless steel when identical test conditions were prevailed.

Gebert et al. [52] has performed a comparative corrosion study of both Mg65Y10Cu15Ag10 and Mg65Y10Cu25 BMGs in a borate buffer solution (pH 8.4) with pure Mg and Mg65Y10Cu25 crystalline alloys. The electrochemical behavior of both amorphous and crystalline Mg65Y10Cu25 alloys was similar, but superior to that of pure Mg. Although, the Mg65Y10Cu15Ag10 BMG exhibited superior corrosion resistance among the other three alloys.

2.2.2. *Corrosion resistance of BMGs in comparison to crystalline alloys*

There is still considerable interest in how the electrochemical properties of metallic glasses and BMGs compare to those of conventional crystalline alloys. Amorphous alloys are believed to exhibit corrosion resistance due to:

- their compositions, which are not constrained by the solubility limits, and they can be allied to elements promoting passivation [47],

- and their lack of microstructural characteristics, such as grain boundaries, dislocations, and precipitates, which commonly serve as sites for the local passive film breakdown entailing pitting.

Comparisons made on amorphous alloys and their crystalline counterparts allowed it possible to determine whether the structural disorder influences their corrosion behavior [55, 56]. Schroeder et al. [55] have shown that the resistance to pitting corrosion of the Zr41.2Ti13.8Cu12.5Ni10Be22.5 BMG was nearly as equivalent as its crystalline counterpart when exposed to 0.5 M NaCl solution. However, the BMG was not more resistant to uniform corrosion in a 0.5 M $NaClO_4$ solution as its crystalline counterpart did. In accordance with this conclusion, Köster et al. [57] showed that there was no significant difference between the polarization behavior of the amorphous and nanocrystalline states of Zr69.5Cu12Ni11Al7.5 in a 0.1 N NaOH. However, Peter et al. [56] have found that the Zr52.5Cu17.9Ni14.6Al10Ti5 BMG exhibited lower corrosion rates and greater resistance to pitting in a 0.6 M NaCl solution relative to their corresponding crystalline alloy of similar composition. Naka et al. [9] reported higher corrosion rates of amorphous Fe70Cr10B20 (ribbon) in a 1 M HCl solution at 303 K relative to the crystalline Fe90Cr10 alloy. Nevertheless, the amorphous Fe50Cr30B20 alloy experienced a lower corrosion rate in 1 M HCl than that of the Fe70Cr30 crystalline alloy. It has been cautioned that a convenient way to make an accurate assessment of the role of structural disorder on the corrosion behavior can only be achieved when the corrosion properties of the amorphous alloy have to be compared to those of a single-phased crystalline alloy of the same composition [58].

When considering the corrosion properties, the effect of the structure must also be addressed. However, the composition of the alloy is more likely to have a much greater influence on the electrochemical behavior of the alloy than if its structure is periodic or in a disordered state [58].

2.3. Tribological and tribocorrosion properties of BMGs

So far, limited literature information is available on the tribocorrosion properties of BMGs. Therefore, a brief overview is given below with respect to the recent major breakthrough in tribological and tribocorrosion performance of BMGs. The evaluation of metallic glasses as potentially applicable in situations where wear and corrosion act simultaneously is still premature.

Among the most important means of characterizing the performance of BMGs, materials scientists opt for the examination of the tribological and/or tribocorrosion resistance of a BMG material. The most studied material systems are zirconium-based bulk metallic glasses (Zr-BMGs) because of their wide use as mechanical components, their simple preparation process, and their excellent properties.

2.3.1. Zr-based BMGs

The tribocorrosion behavior of Zr55Cu30Ni5Al10 BMG sliding against AISI 52100 steel bearing in 3.5% NaCl solution in a pin-on-disk tribometer was evaluated by Ji et al. [59]. The wear volume loss of the Zr55Cu30Ni5Al10 BMG increased as the applied load (up to 20 N), the sliding distance and the sliding velocity (up to 1.5 m s^{-1}) increased. The BMG surface structure was believed to evolve from an amorphous to a pseudo-crystalline nanostructure during sliding wear conditions. Due to the BMG surface crystallization, the wear resistance was improved and the friction coefficient (μ) of the Zr-based BMG-steel tribo-pair decreased from 0.5 to 0.08, but its free corrosion resistance has been lowered. It has also been shown that its degradation mechanisms evolve during sliding-corrosion tests as a function of normal load.

Abrasive wear was the dominant degradation mechanism at a load of 5 N, while adhesive and corrosive wear were predominant for an applied load of 20 N. Compared to AISI 304 stainless steel under similar test conditions, the Zr-based BMG has better tribocorrosion resistance in the NaCl solution but also beneficial in reducing the wear rate of the bearing steel disk. Therefore, the Zr-based BMG could be considered as a good candidate material for tribocorrosion in seawater applications.

2.3.1.1. *Effect of the counter-body material*

A recent study by Tian et al. [60] focused on the tribological properties of Zr-based BMG sliding against polymers, steels, and ceramics at various loads and speeds. Acoustic emission (AE) technology has been used to analyze the evolution of wear.

The coefficients of friction (μ) of the BMG sliding against either steel or ceramic balls were high (about 0.6–0.85) but decreased with increasing normal load and sliding velocity. As the steel balls were more ductile than the ceramic ones, the steel-BMG tribo-pairs generated weaker AE signals and exhibited larger wear rates. The main wear mechanism of the steel-BMG and ceramic-BMG pairs was dominated by a plastic flow of the BMG surface material caused by structural relaxation. The AE signals in a single frictional sequence decrease slightly with the increase of the sliding velocity due to the elastic energy consumed by the structural relaxation. The wear surfaces of Zr-based BMG samples were smoother, without crushing and peeling due of the super-plasticity of these BMGs over crystalline metallic materials.

Zr-based BMG sliding against polymer balls had much lower and more stable μ (0.5–0.65) than Zr-based BMG sliding against steel or ceramic balls. The low resistance to friction has been attributed to interfacial material transfer of polymer layers to the BMG surface during repeated sliding. The BMG sliding against polymer balls exhibited also the highest AE signals among the three types of counter-body materials considered, indicating that abrasive wear dominated the degradation mechanism in polymer-BMG tribo-pairs. Due to the low hardness and strength of the polymer counter-bodies, the BMG surfaces are not worn off. In addition to the abrasive wear mechanism in polymer-BMG tribo-pairs, a predominant process of adhesive wear was observed when the polymer wear debris (layers) were transferred to the BMG surfaces. Both the worn circular contact area and the wear volume loss of the polymer increased as the sliding distance increased. Similarly, plowing and fracture of the asperities increased after AE signals were found to increase in a single friction sequence. The authors call for the potential application of these types of BMGs in areas where tribology plays an important role. These Zr-based BMGs could act as new alternative candidates for tribo-materials (e.g., counter-body materials) over their traditional crystalline ones.

Zhong et al. [61] studied the tribological properties of $Zr41Ti14Cu12.5Ni10Be22.5$ pins sliding against different counter-body materials, namely AISI 5120 steel, AISI 52100 steel, and

Zr41Ti14Cu12.5Ni10Be22.5 amorphous alloys using a pin-on-disk tribometer at room temperature under relatively heavy loads. The μ depended on the counterbody material. The highest value of μ was measured for AISI 5120 steel (0.30) under a normal load of 150 N, whereas the lowest value (0.15) was recorded for the Zr-based BMG disk for a normal applied load of 100 N. The wear weight loss of the crystalline material disk was lower than that of the pins. The opposite result was obtained when the Zr-based BMG was used as the counterbody material. For the three different counterbody materials, the crystalline disks showed obvious piled-up material without severe peeling-off. In addition, microcracks were found on the AISI 5120 steel disk worn surface, whereas the plastic flow has become one of the dominant wear mechanisms on the AISI 52100 steel disk. However, rough detached surface damage and wave-like patterns accompanied by pronounced delamination have been found on the BMG pins when amorphous disks were used as counterpart. Differences in wear mechanisms have been observed between the crystalline materials and the amorphous disks. The wear mechanisms of Zr-based BMG pins included grooves, micro-cracks, peeling-off and vicious flow when they slid against crystalline materials.

2.3.1.2. *Effect of the loading conditions*

The tribological behavior of a Zr52.5Cu17.9Ni14.6Al10Ti5 (at.%) BMG sliding against yttria-stabilized zirconia counterparts was investigated using pin-on-disk in two distinct environments, namely air and argon [62]. It was found that the wear of the Zr-based BMG was reduced by more than 45% due to the removal of oxygen from the test environment at two different loads, namely, 16 and 23 N. The surface wear pins were examined using X-ray diffractometry, differential scanning calorimetry, scanning electron microscopy and optical surface profilometry. A number of abrasive particles and grooves were observed on the worn surface of the pin tested in air, while a relatively smooth worn surface was found for specimens tested in argon. In ambient air, an abrasive wear mechanism dominates the degradation mechanism of the BMG pin, whereas, in argon, an adhesive wear controls the wear process.

The effect of the load on the wear behavior of Zr55Cu30Ni5Al10 BMG under linear reciprocating sliding conditions was investigated by Tao et al. [63]. An increase in the normal load results in a decrease of the μ of the Zr-based BMG (from 0.348 down to 0.226), and an increase of the wear volume loss. The wear mechanism evolves during friction, combining different processes such as welding, adhesive, and abrasive wear. At low loads, adhesive wear governs the degradation wear mechanism of the Zr-based BMG.

2.3.1.3. *Effect of the test medium*

The wear resistance of a Zr61Ti2Cu25Al12 (ZT1) BMG sliding against Si_3N_4 balls in dry (air), and lubricated states (deionized water, and simulated physiological media) in a ball-on-flat contact configuration was investigated by Wang et al. [64]. It has been shown that the wear resistance of ZT1 BMG in air and deionized water was superior to that of Ti6Al4V alloy but inferior to that of 316 L stainless steel and CoCrMo alloy when similar conditions were used. However, in simulated physiological media, such as phosphate buffered solution (PBS), and DMEM + FBS (Dulbecco's modified Eagle medium with 10 vol.% fetal bovine serum), the Zr-based BMG exhibited lower wear resistance than that of Ti6Al4V, 316 L stainless steel, and

CoCrMo. This was likely attributed to its moderate pitting corrosion resistance, and its poor depassivation-repassivation kinetics, as induced by passive film breakdown in the solution containing chloride ions during the tribo-corrosion conditions. The presence of proteins in the test solution had a significant effect on the rapid decrease in the pitting resistance of the Zr-based BMG, which caused greater and severe wear damage. Therefore, improvement in the pitting resistance of the Zr-based BMG in physiological media is necessary if it concerns its possible use as a biomedical implant. In addition, screening of a good material as a counterpart to coupling with Zr-based BMG is an additional key factor in ensuring better wear resistance (e.g., by lowering its wear rate). In dry conditions, abrasive wear was predominant mechanism in the case of ZT1 BMG. Under lubricated sliding conditions with the presence of deionized water, the wear damage caused by abrasion can be mitigated. In a simulated physiological environment with the presence of chloride ions, synergistic effects of abrasive and corrosive wear more likely control the wear process. For the four metals studied, the wear resistance had no distinct correlation with the hardness, while the material with high Young's modulus had better wear resistance.

In another investigation conducted by Hua et al. [65], the tribological behavior of a Zr53Al16Co23.25Ag7.75 BMG in air and phosphate buffer saline (PBS) solution was evaluated using ball-on-disk reciprocating sliding contacts. A biomedical alloy Ti-6Al-4V was used for comparison purposes. The wear resistance of the Zr-Al-Co-Ag BMG sliding in air has been shown to be superior to that of Ti-6Al-4V alloy. Under such dry friction conditions, the wear damage on Zr-based BMG was governed by combined oxidational and adhesive wear processes, whereas the wear mechanisms on the Ti-6Al-4V alloy was mainly driven by conjoint processes of abrasion and adhesion damage. Similarly, under lubricated contact conditions, both oxidational and abrasive wear were the main wear mechanisms of the Ti-6Al-4V alloy in PBS. The Zr-based BMG exhibited lower wear resistance under lubricated sliding conditions (i.e., PBS) over dry friction, but higher than that of Ti-6Al-4V alloy. That low wear resistance was likely attributed to the low pitting corrosion resistance of Zr-Al-Co-Ag BMG in the medium containing chloride ions, and to the synergistic effects of abrasive and corrosive wear during the tribocorrosion tests. The pitting resistance in PBS solution of Zr-Al-Co-Ag BMG was lower than that of Ti-6Al-4V alloy, and was considered a key factor in the tribocorrosion behavior of the Zr-based BMG under the conditions considered.

2.3.2. *Fe-based BMGs*

The sliding wear behavior of a range of rapidly solidified alloy samples based on the composition of Fe68Cr18Mo2B12 (prepared by planar-flow casting) against a cobalt-bonded tungsten carbide counterface was evaluated using a modified crossed-cylinder wear testing rig [66]. A range of microstructures was examined, and the effect on the wear performance was evaluated. Alloys in ribbon forms were investigated in the as-cast amorphous state and after devitrification at various temperatures, and with the bulk-devitrified sample prepared by hot extrusion of the crushed ribbon. Low-temperature crystallization of the ribbon produced a high volume fraction of metastable Fe_3B in a ferritic matrix, whereas high-temperature crystallization produced M_2B and Mo-rich borides, still in a ferrite matrix. The wear results showed that the alloy having the stable M_2B and Mo-rich borides had the best wear performance, whereas that of the amorphous material had the highest wear rate, and the alloy with

Fe_3B precipitates exhibited an intermediate wear behavior. The wear volume was systematically higher for the amorphous alloy than for the corresponding devitrified material. This was believed to be the result of the continuous crystallization of the glassy ribbon on the worn surface. Crystallization produced a microstructure of deformed ferrite containing a uniform dispersion of very fine boride precipitates. The combination of extensive microcracks and the inability of the precipitates to support the load led to an increase in the wear rate.

3. Toward the use of BMGs for tribo-electrochemistry systems in biomedical applications

The unique properties of BMGs make these materials attractive alternatives in biomedical applications. However, the most promising strategies for biomedical applications are quite limited. BMGs have potential biomedical applications as screws due to their toughness and high strength. Moreover, it has been reported that certain BMG compositions have low magnetic susceptibility, which could be advantageous in surgical instrument applications for interventional magnetic resonance imaging (MRI) [67, 68]. The ease of micro-forming and manufacturing of BMGs [69] also lends itself to the production of gears for small, high-powered micromotors that could be of use in arthroscopic tools. Another promising application is the biofunctionalization technology. Biofunctionalization consists of adapting compounds, for example metals, to make them compatible with a biomedical application. Finally, one of the most promising biomedical applications of BMGs is the use in bone fracture fixation and hip arthroplasty. BMGs have a low modulus that is comparable to the modulus of the bone as well as a high strength to withstand the significant forces generated in the skeletal system of the human body.

In particular, the acceptability of BMGs as potential candidates for implants (e.g., load-bearing materials) in the human body is a very interesting and novel topic for a recent field of applications. To be effective, the BMGs should behave like biomaterials for the duration of their use. Above all, they should have sufficient mechanical strength, corrosion and wear resistance to withstand the harsh conditions of the body environment. If this is not achieved, degradation of the implant occurs and secondary effects take place leading to carcinogenicity, hypersensitivity, inflammation, and other complications. Recent investigations have claimed potential tribocorrosion performance of BMGs in simulated body fluids; however, improvements are needed to achieve full functioning.

The tribocorrosion performance of a Zr-based BMG, namely ZrCuAlNi, has recently been evaluated in different media simulating body fluids with and without the presence of proteins [70]. For comparison, a low-alloyed Zr material and a crystalline structure of the same alloy were also studied. It has been shown that the adsorption of the protein on the metal surface modifies both the cathodic and anodic electrochemical patterns and thus changes the prevailing reactions. The ZrCuAlNi BMG alloy in both structures, that is, amorphous and crystalline, did not form any stable passive film and an active dissolution in the anodic region was observed; only the presence of phosphates has generated a small passive plateau. The results also showed that the BMG material with amorphous structure had the largest wear rates in all electrolytes tested (i.e., NaCl, phosphate buffer solution, and phosphate buffer solution with protein), whereas the crystalline structure of the BMG showed negligible wear rates. Two different tribocorrosion mechanisms were observed in the ZrCuAlNi BMG alloy. The amorphous structure suffered from wear-accelerated corrosion generating larger wear rates due to galvanic coupling effects. The crystalline structure did not suffer from wear accelerated corrosion (no galvanic coupling effects due to lack of passivity) and thus lower wear rates were measured. The amorphous BMG material generated a large amount of wear debris during tribocorrosion tests, which act as very abrasive particles, thereby entailing the wear of the counterpart material (alumina). This was due to the formation of grooves and the growth/breakdown sequences of an amorphous oxide film and the high availability of Zr in the material. Crystalline BMG material generated low amount of wear debris, which could be attributed to recrystallization (nano-crystallization), and the presence of a new intermetallic phase (Zr_2Cu), promoting an improvement of its mechanical properties under tribocorrosion conditions.

In another comparable study [71], using a reciprocating ball-on-disk tribometer equipped with an electrochemical cell, the degradation mechanisms due to mechanical wear and tribocorrosion of Zr-based BMGs, with nominal compositions of Zr55Cu30Ni5Al10 and Zr65Cu18Ni7Al10 used as potential load-bearing implant materials, were examined. The Zr65-BMG with a higher Zr content showed an increase in plasticity but a reduction in its wear resistance during sliding was noticed. Both BMGs experienced abrasive wear mechanisms after dry sliding wear test conditions (under applied load of 2 N). The worn surface had a higher hardness and no wear-induced recrystallization was confirmed *via* FIB cross-section analysis. Therefore, the hardening mechanism was due to the free-volume annihilation under suppressed shear deformation. The more passive nature of the Zr65-BMG had consequently a negative influence on its tribocorrosion resistance since wear-accelerated corrosion speeded-up its governing degradation mechanism. It has been shown that a galvanic coupling established between the depassivated wear track and the surrounding passive area was behind the main degradation mechanism of passive Zr65-BMG subjected to the tribocorrosion environment. In order to apply the BMGs for load-bearing biomedical devices, it is required to find out a balance between their wear-controlled mechanisms, namely wear-accelerated corrosion and static corrosion resistance.

Author details

Abdenacer Berradja

Address all correspondence to: a.berradja@gmail.com

MTM Department, K.U. Leuven, Leuven, Belgium

References

[1] Kramer J. The amorphous state of metals. Zeitschrift für Physik. 1937;**106**:675-691

[2] Brenner A, Couch DE, Williams EK. Electrodeposition of alloys of phosphorus with nickel or cobalt. J. Research of the National Bureau of Standards. 1950;**44**:109

[3] Klement W, Willens RH, Duwez P. Non-crystalline structure in solidified gold-silicon alloys. Nature. 1960;**187**:869-870

[4] Naka M, Hashimoto K, Masumoto T. Corrosion resistance of amorphous iron alloys with chromium. J. Japan Institute of Metals. 1974;**38**(9):835-841

[5] Hays CC, Kim CP, Johnson WL. Microstructure controlled shear band pattern formation and enhanced plasticity of bulk metallic glasses containing in situ formed ductile phase dendrite dispersions. Physical Review Letters. 2000;**84**:2901-2904

[6] Inoue A, Kimura HM, Sasamori K. Masumoto T. Ultrahigh strength of rapidly solidified Al-(96-x), Cr-3, Ce-1, Co-x (x = 1, 1.5 and 2%) alloys containing an icosahedral phase as a main component. Materials Transactions. 1994;**35**:85

[7] Flores KM, Dauskardt RH. Enhanced toughness due to stable crack tip damage zones in bulk metallic glass. Scripta Materialia. 1999;**41**:937-943

[8] Hashimoto K, Naka M, Masumoto T. Scientific Report of the Research Institutes of Tohoku University, A-26. Tohoku University; 1976. p. 48

[9] Naka M, Hashimoto K, Masumoto T. Corrosion resistivity of amorphous Fe alloys containing Cr. Journal of the Japan Institute of Metals. 1974;**38**(9):835-841

[10] Masumoto T, Hashimoto K. Chemical properties of amorphous metals. Annual Review of Materials Science. 1978;**8**:215-233

[11] Naka M, Hashimoto K, Masumoto T. High Corrosion Resistance of Chromium-Bearing Amorphous Iron Alloys in Neutral and Acidic Solutions Containing Chloride. Corrosion. 1976;**32**(4):146-152

[12] Diegle RB. Localized Corrosion of Amorphous Fe-Ni-Cr-P-B Alloys. Corrosion. 1979; **35**(6):250-258

[13] Pang SJ, Zhang T, Asami K, Inoue A. Bulk glassy Fe-Cr-Mo-C-B alloys with high corrosion resistance. Corrosion Science. 2002;**44**(8):1847-1856

[14] Asami K, Pang SJ, Zhang T, Inoue A. Preparation and corrosion resistance of Fe-Cr-Mo-C-B-P bulk glassy alloys. Journal of the Electrochemical Society. 2002;**149**(8):B366-B369

[15] Naka M, Hashimoto K, Masumoto T. High corrosion resistance of amorphous Fe-Mo and Fe-W alloys in HCl. J. Non-Crystalline Solids. 1978;**30**(1):61-65

[16] Asami K, Naka M, Hashimoto K, Masumoto T. Journal of the Electrochemical Society. 1980;**127**:2130

[17] Pang SJ, Zhang T, Asami K, Inoue A. Formation of bulk glassy Fe75−x−yCrxMoyC15B10 alloys and their corrosion behavior, J. Materials Research. 2002;**17**(3):701-704

[18] Naka M, Hashimoto K, Masumoto T. Change in corrosion behavior of amorphous Fe-P-C alloys by alloying with various metallic elements. J. Non-Crystalline Solids. 1979;**31**(3):355-365

[19] Wang R. Fall Meeting. Pittsburgh, PA: The Metallurgical Society; 1980

[20] Shimamura K, Kawashima A, Asami K, Hashimoto K. Corrosion behavior of amorphous nickel-valve metal-alloys in boiling concentrated nitric and hydrochloric acids. Science Reports of the Research Institutes, Tohoku University. 1986;**33**(1):196-210

[21] Mitsuhashi A, Asami K, Kawashima A, Hashimoto K. The corrosion behavior of amorphous nickel base alloys in a hot concentrated phosphoric acid. Corrosion Science. 1987;**27**(9):957-970

[22] Lee HJ, Akiyama E, Habazaki H, Kawashima A, Asami K, Hashimoto K. The effect of phosphorus addition on the corrosion behavior of amorphous Ni-30Ta-P alloys in 12 M HCl. Corrosion Science. 1995;**37**(2):321-330

[23] Diegle RB. Crevice Corrosion of Glassy Fe-Ni-Cr-P-B Alloys. Corrosion. 1980;**36**(7):362-368

[24] Pang SJ, Shek CH, Zhang T, Asami K, Inoue A. Corrosion behavior of glassy Ni55Co5Nb20Ti10Zr10 alloy in 1N HCl solution studied by potentiostatic polarization and XPS. Corrosion Science. 2006;**48**(3):625-633

[25] Habazaki H, Sato T, Kawashima A, Asami K, Hashimoto K. Preparation of corrosion-resistant amorphous Ni-Cr-P-B bulk alloys containing molybdenum and tantalum. Materials Science and Engineering A. 2001;**304-306**:696-700

[26] Pang SJ, Zhang T, Asami K, Inoue A. Bulk glassy Ni(Co-)Nb-Ti-Zr alloys with high corrosion resistance and high strength. Materials Science and Engineering A. 2004;**375-377**:368-371

[27] Hashimoto K, Kasaya M, Asami K, Masumoto T. Corrosion Engineering. 1977;**26**:445

[28] Latanision RM, Turn JC, Compeau CR. In: Proceedings of the Third International Conference on Mechanical Behavior of Metals. Vol. 2. 1979. p. 475

[29] Pang SJ, Zhang T, Asami K, Inoue A. Synthesis of Fe–Cr–Mo–C–B–P bulk metallic glasses with high corrosion resistance. Acta Mater. 2002;**50**(3):489-497

[30] Lin XH, Johnson WL. Formation of Ti-Zr-Cu-Ni bulk metallic glasses. Journal of Applied Physics. 1995;**78**(11):6514-6519

[31] Inoue A, Zhang W, Zhang T, Kurosaka K. High-strength Cu-based bulk glassy alloys in Cu-Zr-Ti and Cu-Hf-Ti ternary systems. Acta Materialia. 2001;**49**(14):2645-2652

[32] Inoue A, Zhang W, Zhang T, Kurosaka K. Thermal and mechanical properties of Cu-based Cu-Zr-Ti bulk glassy alloys. Materials Transactions. 2001;**42**(6):1149-1151

[33] Asami K, Qin CL, Zhang T, Inoue A. Effect of additional elements on the corrosion behavior of a Cu-Zr-Ti bulk metallic glass. Materials Science and Engineering A. 2004;**375-377**:235-239

[34] Liu B, Liu L. The effect of microalloying on thermal stability and corrosion resistance of Cu-based bulk metallic glasses. Materials Science and Engineering A. 2006;**415**(1-2): 286-290

[35] Qin CL, Asami K, Zhang T, Zhang W, Inoue A. Corrosion behavior of Cu-Zr-Ti-Nb bulk glassy alloys. Materials Transactions. 2003;**44**(4):749-753

[36] Qin CL, Zhang W, Asami K, Ohtsu N, Inoue A. Glass formation, corrosion behavior and mechanical properties of bulk glassy Cu-Hf-Ti-Nb alloys. Acta Materialia. 2005;**53**(14):3903-3911

[37] Liu L, Liu B. Influence of the micro-addition of Mo on glass forming ability and corrosion resistance of Cu-based bulk metallic glasses. Electrochimica Acta. 2006;**51**(18):3724-3730

[38] Tam MK, Pang SJ, Shek CH. Effects of niobium on thermal stability and corrosion behavior of glassy Cu-Zr-Al-Nb alloys. Journal of Physics and Chemistry of Solids. 2006;**67**(4):762-766

[39] Qin CL, Zhang W, Kimura H, Asami K, Inoue A. New Cu-Zr-Al-Nb bulk glassy alloys with high corrosion resistance. Materials Transactions. 2004;**45**(6):1958-1961

[40] Inoue A, Zhang T, Nishiyama N, Ohba K, Masumoto T. Preparation of 16 mm diameter rod of amorphous Zr65Al7.5Ni10Cu17.5 alloy. Materials Transactions. 1993;**34**(12):1234-1237

[41] Peker A, Johnson WL. A highly processable metallic-glass—Zr41.2Ti13.8Cu12.5Ni10.0Be22.5. Applied Physics Letters. 1993;**63**(17):2342-2344

[42] Asami K, Habazaki H, Inoue A, Hashimoto K. Recent development of highly corrosion resistant bulk glassy alloys. Materials Science Forum. 2005;**502**:225-230

[43] Qin FX, Zhang HF, Deng YF, Ding BZ, Hu ZQ. Corrosion resistance of Zr based bulk amorphous alloys containing Pd. Journal of Alloys and Compounds. 2004;**375**(1-2):318-323

[44] Pang SJ, Zhang T, Asami K, Inoue A. Formation, corrosion behavior, and mechanical properties of bulk glassy Zr-Al-Co-Nb alloys. Journal of Materials Research. 2003;**18**(7):1652-1658

[45] Raju VR, Kühn U, Wolff U, Schneider F, Eckert J, Reiche R, Gebert A. Corrosion behaviour of Zr-based bulk glass-forming alloys containing Nb or Ti. Materials Letters. 2002;**57**(1):173-177

[46] Liu L, Qiu CL, Zou H, Chan KC. The effect of the microalloying of Hf on the corrosion behavior of ZrCuNiAl bulk metallic glass. Journal of Alloys and Compounds. 2005;**399**(1-2):144-148

[47] Gebert A, Mummert K, Eckert J, Schultz L, Inoue A. Electrochemical investigations on the bulk glass forming Zr55Cu30Al10Ni5 alloy. Materials and Corrosion. 1997;**48**(5):293-297

[48] Mudali UK, Baunack S, Eckert J, Schultz L, Gebert A. Pitting corrosion of bulk glass-forming zirconium-based alloys. Journal of Alloys and Compounds. 2004;**377**:290-297

[49] Hiromoto S, Tsai AP, Sumita M, Hanawa T. Effect of chloride ion on the anodic polarization behavior of the Zr65Al7.5Ni10Cu17.5 amorphous alloy in phosphate buffered solution. Corrosion Science. 2000;**42**(9):1651-1660

[50] Gebert A, Buchholz K, El-Aziz AM, Eckert J. Hot water corrosion behaviour of Zr-Cu-Al-Ni bulk metallic glass. Materials Science and Engineering A. 2001;**316**(1-2):60-65

[51] Morrison ML, Buchanan RA, Senkov ON, Miracle DB, Liaw PK. Electrochemical behavior of Ca-based bulk metallic glasses. Metallurgical and Materials Transactions A. 2006;**37**(4):1239-1245

[52] Gebert A, Subba Rao RV, Wolff U, Baunack S, Eckert J, Schultz L. Corrosion behaviour of the Mg65Y10Cu15Ag10 bulk metallic glass. Materials Science and Engineering A. 2004;**375-377**:280-284

[53] Morrison ML, Buchanan RA, Peker A, Liaw PK, Horton JA. Electrochemical behavior of a Ti-based bulk metallic glass. Journal of Non-Crystalline Solids. 2007;**353**(22-23): 2115-2124

[54] Inoue A, Masumoto T. Mg-based amorphous alloys. Materials Science and Engineering A. 1993;**173**(1-2):1-8

[55] Schroeder V, Gilbert CJ, Ritchie RO. Comparison of the corrosion behavior of a bulk amorphous metal, Zr41.2Ti13.8Cu12.5Ni10Be22, with its crystallized form. Scripta Materialia. 1998;**38**(10):1481-1485

[56] Peter WH, Buchanan RA, Liu CT, Liaw PK, Morrison ML, Carmichael CA Jr, Wright JL. Localized corrosion behavior of a zirconium-based bulk metallic glass relative to its crystalline state. Intermetallics. 2002;**10**(11-12):1157-1162

[57] Köster U, Zander D, Triwikantoro RA, Jastrow L. Environmental properties of Zr-based metallic glasses and nanocrystalline alloys. Scripta Materialia. 2001;**44**(8-9):1649-1654

[58] Archer MD, Corke CC, Harji BH. The electrochemical properties of metallic glasses. Electrochimica Acta. 1987;**32**(1):13-26

[59] Ji X, Hu B, Li Y, Wang S. Sliding tribocorrosion behavior of bulk metallic glass against bearing steel in 3.5% NaCl solution. Tribology International. 2015;**91**:214-220

[60] Tian P, Khun NW, Tor SB, Liu E, Tian Y. Tribological behavior of Zr-based bulk metallic glass sliding against polymer, ceramic, and metal materials. Intermetallics. 2015;**61**:1-8

[61] Zhong H, Chen J, Dai LY, Yue Y, Wang BA, Zhang XY, Ma MZ, Liu RP. Effect of counterpart material on the tribological properties of Zr-based bulk metallic glass under relatively heavy loads wear. 2015;**346**:22-28

[62] Wu H, Baker I, Liu Y, Wu X-l. Dry sliding tribological behavior of Zr-based bulk metallic glass. Transactions of Nonferrous Metals Society of China. 2012;**22**(3):585-589

[63] Tao PJ, Yang YZ, Bai XJ, Ru Q, Xie ZW. Effect of load on linear reciprocating sliding friction and wear behavior in Zr-based bulk metallic glass. Advanced Materials Research. 2011;**152-153**:1905-1908

[64] Wang Y, Shi LL, Duan DL, Li S, Xu J. Tribological properties of Zr61Ti2Cu25Al12 bulk metallic glass under simulated physiological conditions. Materials Science & Engineering. C, Materials for Biological Applications. 2014;**37**:292-304

[65] Hua N, Zheng Z, Fang H, Ye X, Lin C, Li G, Wang W, Chen W, Zhang T. Dry and lubricated tribological behavior of a Ni- and Cu-free Zr-based bulk metallic glass. Journal of Non-Crystalline Solids. 2015;**426**:63-71

[66] Anis M, Rainforth WM, Davies HA. Wear behaviour of rapidly solidified Fe68Cr18Mo2B12 alloys. Wear. 1994;**172**(2):135-145

[67] Seifert V, Zimmermann M, Trantakis C, Vitzthum HE, Kuhnel K, Raabe A, Bootz F, Schneider JP, Schmidt F, Dietrich J. Open MRI-guided neurosurgery. Acta Neurochirurgica. 1999;**141**:455-464

[68] Horton JA, Parsell DE. Biomedical potential of a zirconium-based bulk metallic glass. MRS Proceedings In: Symposium CC- Supercooled Liquids, Glass Transition and Bulk Metallic Glasses; Vol. 754, CC1.5; 2002. DOI: 10.1557/PROC-754-CC1.5

[69] Saotome Y, Hatori T, Zhang T, Inoue A. Superplastic micro/nano-formability of La-60, Al-20, Ni-10, Co-5, Cu-5 amorphous alloy in supercooled liquid state. Materials Science and Engineering A. 2001;**304-306**:716-720

[70] Espallargas N, Aune RE, Torres C, Papageorgiou N, Muñoz AI. Bulk metallic glasses (BMG) for biomedical applications—A tribocorrosion investigation of Zr55Cu30Ni5Al10 in simulated body fluid. Wear. 2013;**301**(1-2):271-279

[71] Zhao GH, Aune RE, Mao H, Espallargas N. Degradation of Zr-based bulk metallic glasses used in load-bearing implants: A tribocorrosion appraisal. J. Mechanical Behavior of Biomedical Materials. 2015;**60**:56-67